Playing to Win

DIGITAL GAME STUDIES

Robert Alan Brookey and David J. Gunkel, editors

INDIANA UNIVERSITY PRESS

Bloomington & Indianapolis

PLAYING TO WIN

Sports, Video Games, and the Culture of Play

EDITED BY
Robert Alan Brookey
and Thomas P. Oates

This book is a publication of

INDIANA UNIVERSITY PRESS
Office of Scholarly Publishing
Herman B Wells Library 350
1320 East 10th Street
Bloomington, Indiana 47405 USA

iupress.indiana.edu

Telephone   800-842-6796
Fax   812-855-7931

© 2015 by Indiana University Press

∞ The paper used in this publication
meets the minimum requirements of
the American National Standard for
Information Sciences–Permanence of
Paper for Printed Library Materials,
ANSI Z39.48–1992.

*Manufactured in the
United States of America*

*Library of Congress
Cataloging-in-Publication Data*

Cataloging information is available from
the Library of Congress.

ISBN 978-0-253-01499-3 (cloth)
ISBN 978-0-253-01502-0 (paperback)
ISBN 978-0-253-01505-1 (ebook)

1 2 3 4 5   20 19 18 17 16 15

Contents

Playing to Win

Introduction

Thomas P. Oates and Robert Alan Brookey

PERHAPS ONE OF THE MOST PERSISTENT LEGENDS OF THE
early video game industry involves the installation of a prototype of
the *Pong* game in a Sunnyvale, California, bar named Andy Capp's in
September 1972.[1] Two weeks after the game was installed, Atari engineer
Al Alcorn got a call from the bar manager, complaining that the game
was broken and requesting that it be hauled off the premises. When
Alcorn went to investigate the problem, he discovered the machine was
jammed and overflowing with quarters. This story certainly has all the
trappings of a corporate myth, but it offers an event marking the beginning of the rise of Atari as a leader in the video game industry. Given the
significantly diminished status of Atari and its recent bankruptcy, it is
important to remember Atari's former prominence. In other words, this
story about Andy Capp's also marks an important moment for the video
game industry in general.

Yet there is another important point about this event that is often
overlooked: if *Pong* was one of the first successful commercial video
games, then one of the first successful video games was a sports simulation game. We can anticipate the snickering this observation might
inspire. After all, *Pong* was incredibly primitive, and table tennis (or Ping-
Pong) enjoys a dubious place in the pantheon of sports – it's right up
there with badminton and croquet. We could counter that Ping-Pong
was used to open a diplomatic relationship between the United States
and China in 1971, just over a year prior to the installation of the *Pong*
machine at Andy Capp's. Our point, however, is not about the legitimacy of Ping-Pong as a sport, but rather the importance of sports to the

emergence of the video game industry. At about the time *Pong* was released as an arcade game, Magnavox unveiled its Odyssey home gaming
system. From its earliest incarnations, mass-market video gaming has
simulated popular sports. Atari's 1972 breakthrough success, *Pong,* was
a table-tennis simulation, while competitors sought to replicate games
such as tennis, hockey, baseball, and football. Like *Pong,* the Odyssey
had very abstract graphics that were augmented with plastic overlays
that could be placed on the screen of the television set. These overlays
were designed to simulate various games, including tennis, hockey, football, and table tennis.[2] In fact, Magnavox and Atari became embroiled
in a lawsuit over the rights to simulate Ping-Pong as a video game; the
suit was finally settled out of court, much to Atari's favor.[3] A few years
later, Mattel would launch the first handheld LED gaming systems that
included versions of football, baseball, and basketball.[4] Central to the
initial success of video gaming, sports simulation games have held an
important place in the history of every sector of the video game market
since, including the arcade, console, and handheld markets.

We are far removed from the seventies, and in the intervening years
the technological sophistication of video games has evolved far beyond
the offerings of *Pong* and its brethren. The video game industry has not
only developed a great number of sports simulation games over the years,
but also built a strong relationship with the brands and franchises in the
sports industry. To put this in perspective, it might be helpful to consider
some numbers. In the Entertainment Software Association 2011 report,
U.S. households in 2010 reportedly spent $15.9 billion on video game
software, and sports games were the second most popular genre, with
16.3 percent of the market. In that year, almost $2.6 billion was spent on
sports games in the United States alone.[5] In addition, of the top-ten video
games in 2010, *Madden NFL 11* was ranked second, and *NBA 2K11* was
ranked tenth. Therefore, the overlap between the video game and sports
industries is significant, not only in terms of the actual game sales, but
also in the sport brands they represent.

In many respects, the terrain of contemporary sport is suffused
by video gaming, and the boundaries separating the two spheres have
blurred significantly. The top-selling sports-themed video games are
packaged with the images of players prominently displayed on the cov-

ers, and release parties and other promotional events routinely include the presence of star athletes from the present and recent past. The sporting press constantly reports the wild popularity of video game sports simulations among professional athletes, while sports highlight and analysis programs frequently employ video game simulations as pregame analysis and to help predict outcomes. In such games, sporting celebrities are reproduced in digital form, ascribed particular abilities, and placed in competition with others. The terms of athletic competition also suffuse the emerging genre of e-sports, professional leagues in which gamers compete with one another for fame and fortune akin (if not in scale) to that enjoyed by sporting celebrities. Meanwhile, video gaming has become an unlikely site for physical fitness initiatives. The Nintendo Wii system brought "exergaming" to the video game market, creating marketing opportunities and prompting interest among policy wonks.

This overlap between video games and sports provides some interesting opportunities for critical engagement, and this book is devoted to studying the points of convergence for these two industries. But before we turn our attention to the importance of sports games and their relationship to sports culture, we ought to briefly review how the study of video games and the study of sports have evolved over the past few years.

Although video games have been around for more than four decades, they have seldom been the concern of media scholars. Perhaps this is because for much of their existence, video games were not considered to be an artistically legitimate media form. Those media scholars who did show concern for video games focused on the effects of video games, specifically the negative effects they might have on children.[6] Although media effects scholars have continued this work, another type of video game scholar began to emerge about a decade ago. These were scholars trained in critical methods and cultural theory, and they were interested in integrating how video games operated as legitimate forms of social, political, and cultural expression. In particular, scholars interested in critical and cultural studies began to write and publish on video games, and they approached video games from a variety of perspectives. These studies have looked at how video games reflect other narrative forms, reflect cinema, and represent gender.[7]

Video games began to attract these cultural scholars in part because video games became more complex. From the early technologically primitive and graphically abstract sports simulations, advancements in video game technology gradually developed better graphics and more complex narratives. It is important to note that video games, by and large, are computer programs and that video game consoles are basically computers. These consoles take digital information and use it to render images and movement on the video screen. Just as computers advanced to process more data and became more capable of handling audio and video, so too did video game consoles. Video games thus became more visually realistic and dynamic and more narratively complex. Consequently, these technological advances allow for games that were more culturally expressive than earlier games and therefore more attractive to cultural and critical scholars.

What has emerged is a vital new academic field, interdisciplinary in nature, and broad in its heuristic reach. For example, massively multiplayer online role playing games (MMORPGS) have been examined by Edward Castronova as virtual worlds with realistic economies, whereas T. L Taylor has analyzed the unique social communities that these games produce. Ian Bogost has looked at the rhetorical properties of video games and how such games have been used politically and commercially. Stephen Kline, Nick Dyer-Witheford, and Greig De Peuter have authored a political economy critique of the video game industry.[8] In addition, scholars have critiqued the way video games and virtual environments represent gender and sexuality.[9] In spite of this broadening range of critical interest in video games, however, analyses of sport cultures are conspicuously scarce. This scarcity is surprising because of the important place sports games enjoy in the video game industries and the growing importance of sports culture as a field of academic study. Yet, to cite one example, over the past six years, *Games and Culture,* an important journal in the video games studies literature, has directed almost no attention to sports games. By way of contrast, MMORPGS have been given a great deal of attention, and an entire issue of *Games and Culture* was devoted to the game *World of Warcraft.* There are exceptions, such as Halverson and Halverson's essay on fantasy baseball, but the exceptions are few and far between. More broad accounts of video gaming, such as both editions of *The Video Game Theory Reader,* edited by Mark Wolf and Bernard Perron, do not have any chapters devoted to sports video games. In the now seminal book *The Medium of the Video Game,* Mark J. P. Wolf acknowledges "sports" games as a genre, but he suggests that there is little, if anything, that distinguishes them from other "adaptation" games.[10]

It is unlikely that there is a single simple, definitive reason video games studies have not embraced sports games, but the amount of literature on the subject stands in sharp contrast to the prominence of sports video games. Similarly, media-centered studies of sport have only rarely and fleetingly engaged video games as a lens through which critical theories of sport might be developed. Despite a well-developed body of work exploring sport through television, the printed press, and to a lesser degree film, sport studies has not adequately addressed how new digital technologies and new networks and modes of engagement are changing the cultural work of contemporary mediated sport.[11]

This dearth of scholarship on both sides is especially glaring given the growing importance of these games in promoting the professional sports leagues they simulate and the increasingly prominent convergence between video gaming and sport. To begin addressing this gap, we need to consider not just commercial factors, but the cultural forces with which they are bound up. We will start by discussing how the convergence of these two industries is facilitated by an alignment of licensing practices. We will then consider how this convergence also hinges on labor relations that are inflected with an important cultural and social valence: masculinity.

LICENSES TO THRILL

When considering the convergence of two industries, it is important to note how their business practices align. The video game and sports industries both generate a good deal of revenue from licensing practices. Some of these practices are different, but some align quite nicely.

When home video game consoles switched from hardwired to module games, a particular business model came to dominate the industry, one often referred to as the "razor/razorblades" model.[12] This model, as the name suggests, draws on the practice of selling razors at a relatively low price, and as a loss leader, with revenue generated from the sale of replacement razorblades. The fact that these razors and blades are sometimes marketed by sports figures is merely a coincidence at this point, but one still worthy of note. Where the video game industry is concerned, revenue is not usually generated by the sale of gaming consoles. In fact, when a new generation of console is released, its retail price is often below the cost of production.

Profits are generated through the licensing of video game software. Each console basically functions as a computer operating system, and games must be designed to run on the system. Game publishers pay the console manufacture a licensing fee, which provides them the documentation for the game console operating system and the rights to the logo for the system. When a game is published it can be available for different platforms, and while the packaging will be similar, the different console brand logos (Xbox 360, PS3, Wii) will always be prominently displayed. Of course, these brand logos communicate to consumers whether the game will play on the consoles they own, but they also signify a contractual commercial arrangement.[13]

In addition to the practices of hardware and software licensing, the video game industry has been heavily involved in practices associated with licensed content from other industries. For example, Robert Alan Brookey has written extensively about the practices of producing video games based on popular films. This is a fairly common practice for certain kinds of films, and it is one easily facilitated by related practices in the film and video game industries. For example, both industries depended heavily on generic constraints (first-person shooters and role-playing games in the video game industry and action/adventure and computer-generated animation in the film industry) and franchises (*Call of Duty* and the *Grand Theft Auto* games and the *Spiderman* and *The Lord of the Rings* films). Alexis Blanchet has conducted quantitative, longitudinal analysis of this practices and has found that film-to-game licensing has been in practice for almost forty years and has become progressively more common over the years.[14]

Of course, these practices are not limited to film, nor are they unilateral. Many television shows, ranging from *The Simpsons* to *The Walking*

Dead, have had video game spin-offs. More recently, *Defiance* premiered on the SyFy network only a few weeks after an online video game based on the series was released. In addition, novelizations of video games are published, and Blizzard has published books based on their *World of Warcraft* and *Starcraft* games. Video games also become the source of many licensed ancillary products, including T-shirts, ball caps, and action figures. In other words, licensing practices are integrated into many levels of the video game industry, and those practices have been in existence for quite some time.[15]

For sports organizations, licensing is a major source of revenue. Arrangements with sports clothing companies such as Nike and Adidas provide millions of dollars in revenue to professional leagues and individual franchises. In the U.S., college sports programs also benefit from licensing agreements with clothing manufacturers. Sometimes these licensing deals involve legally questionable, anticompetitive practices. Exploiting antitrust exemptions, sports franchises have routinely engaged in activities that would otherwise be found as anticompetitive (imposing salary caps, for example). Sports leagues have used the exemption to negotiate as a cartel rather than as individual teams, driving up the price for their intellectual properties. However, tenuous claims to antitrust exemptions have met with increasing skepticism by the courts. In 2001, for example, NFL Properties negotiated an exclusive ten-year deal with clothing manufacturer Reebok, and prices more than doubled until 2010, when the Supreme Court ruled that the agreement violated the Sherman Anti-Trust Act.

With respect to video gaming, exclusivity has been an important strategy for top developers – especially EA Sports. In 2004, facing growing competition for 2K Sports' National Football League simulation, EA Sports lobbied hard and paid big for an exclusive license with the NFL and the NFL Players Association (NFLPA). This arrangement immediately ended 2K's threat, leaving *Madden* as the sole NFL simulation; therefore, the arrangement was also anticompetitive. Following news of the five-year deal worth three hundred million dollars, Take Two (2K Sports' parent company) released a statement that expressed the view that the arrangement was "a tremendous disservice to the consumers and sports fans whose funds ultimately support the NFL, by limiting their choices, curbing creativity and almost certainly leading to higher game prices."[16] Nevertheless, it has proved lucrative for the partners, and although the exclusivity deal was suppose to expire in 2013, there are indications that undisclosed parts of the deal will continue for a few years.[17]

EA Sports negotiated a similar exclusivity deal with the governing body for U.S. college sports, the National Collegiate Athletic Association (NCAA). This exclusivity deal, which applied to college football and basketball simulations, was the subject of a class-action legal challenge from video game purchasers. The suit ended with EA agreeing to a settlement worth twenty-seven million dollars for violating California consumer protection laws. The settlement left the *Madden NFL's* exclusivity deal unresolved. The fact that the *Madden* exclusivity deal was also negotiated with the NFLPA likely protects the league from antitrust charges. Additionally, both EA Sports and the NFL have expressed deep satisfaction with their current arrangement, and EA's period of exclusivity has left it far ahead of the competition, including 2K, whose pro football simulation was suspended following EA Sports' exclusivity deal with the NFL.[18]

In spite of the tenuous legality of some of these licensing agreements, particularly those between sports franchises and game manufacturers, these agreements are proving to be quite profitable. EA Sports' agreements have allowed it to now extend those licensed games into mobile platforms, so cell phones that are produced by certain manufacturers (such as Motorola), and are preloaded with certain operating systems (such as Android), come preloaded with some of their games.[19] These instances create a trifecta of brand equity, where EA, the mobile phone brand, and the sports franchise all enjoy these visible associations. These brand associations yield advantages above and beyond the actual revenue generated through licensing practices. Now consumers carry with them mobile applications that also function as advertisements for these brands.

Clearly, then, the alignment of sports and video games is facilitated, in no small way, by a myriad of licensing practices. And although some of these practices are questionable, the collaboration and collusion of these two industries continue unabated. In addition to these practices, however, it is also important to note the labor practices of the video game and sports industries and how those practices reflect very similar social and cultural values.

MEN AT WORK

Mediated sport has long facilitated the construction of heroic versions of masculinity. Indeed, many of the most popular modern sports were organized by cultural leaders as a way to confront shifts in economic, cultural, and political life. By cultivating a "muscular Christianity," it was believed that sport could serve as a fortifying practice for white men, as well as a visible symbol of their fitness for leadership. Since then, gender, sexual, and racial relations have changed in important ways, and with it sport has produced new versions of exemplary masculinity, employing and exploiting emerging new media technologies to deepen the pleasure and intensity for subtly shifting audiences.

A number of scholars have examined Western sport, exploring how the new media technologies have changed representational practices that construct masculine icons. For example, Michael Oriard describes how at the turn of the twentieth century, a new emphasis on human interest and the widespread use of images in newspapers helped bring the emerging sport of football to national consciousness and in the process crafted a modern and resonant version of valorized masculinity. Margaret Morse has argued that slow-motion replays on television had a profound effect on popular representations of hypermasculinity. For Morse, slow motion reshaped the "strong cultural inhibition against the look at the male body" prevailing in dominant practices such as football. By focusing attention on the body and disassembling it for review, replay presents male bodies for "erotic display," while also celebrating the male body and its capacities. David Theo Goldberg and David Nylund have separately traced the ways that hegemonic masculinity and whiteness found a novel and affecting new outlet in the explosion of talk radio during the 1990s and 2000s. For Goldberg, sports talk radio quickly became "a leading forum for expressing White maleness," while also marketing a claim to "color-blindness." Nylund found similar themes in his study of sports talk radio personality Jim Rome and concluded that the program was an important cultural site where hegemonic forms of masculinity can be negotiated to meet a changing set of challenges.[20]

Masculinity has a defining construct for the video game industry as well. Although the video game industry has always primarily catered to males, Nintendo locked down the practice of targeting this market when the company single-handedly resurrected the industry in the mid-1980s. Nintendo narrowly focused on a market of young men, and that demographic has continued to dominate the video game market, although women are increasing their share.[21] This dominance is manifest throughout the industry and can be seen in the workforce of the industry and the marketing efforts to address an audience of consumers. For example, when *Dead Space* 2 was released, part of the promotional campaign included a viral video entitled "Your Mom Hates *Dead Space 2*," suggesting that women's rejection of a video game is its greatest recommendation.

Although video gaming has often been positioned as the polar opposite of physical athleticism, the activity usually involves some vicarious association with masculine physical performance. Video game avatars are often capable of extraordinary feats of strength and agility, but those feats are accomplished only if the player successfully manipulates the avatar through the controls. Even when the player uses a female avatar, such as Jade in *Mortal Kombat* or Lara Croft in *Tomb Raider,* they are participating in physically demanding, sometimes violent, activities. In effect, these masculine performances are by proxy and ultimately depend on the skill of the player. Consequently, the decidedly nonathletic activity of sitting on a couch and tapping buttons can become an expression of masculine physical superiority.

This is particularly true when it comes to sports simulation video games, because the gamer often gets to play using an avatar of an actual professional athlete. Again, by proxy, the achievements of the virtual athletes in the game also become the achievements of the gamer, and the gamer experiences the performance of masculinity vicariously. However, the use of actual player likenesses is not without controversy. Currently, the NCAA is facing a significant class-action lawsuit led by former University of California–Los Angeles (UCLA) basketball player

Ed O'Bannon.[22] Under the suit, O'Bannon and his fellow plaintiffs are challenging the NCAA's right to use their likenesses in video games and other profitable products without extending any compensation to the players themselves. The suit has yet to be settled, although the NCAA failed in federal court to have the case thrown out and recently ended its successful *NCAA Football* series in the face of this legal pressure (though a new incarnation of the game is rumored). No matter how it might be settled, the suit reveals another important similarity between the video game and sports industries in the way they both commodify labor.

In pursuit of the hegemonic status that sports heroism affords, many young men are willing to subject themselves to years of low pay and stringent working conditions. This tendency is reflected in the video game industry, where similar sacrifices are made in pursuit of status. For many young men, a career in professional sports would be a "dream job." For many others for whom a professional sports career can be only a dream, a job in the video game industry is a "dream job." The various video game companies know that they have access to a vast pool of young and eager talent willing to work very, very long hours, sometimes for smaller salaries than those offered by other media industries.[23] Consequently, it is common for video game companies to make exorbitant demands of their employees because they know they can.

One notorious practice, called "crunch time," occurs when game development must be accelerated to meet release deadlines. During crunch time, employees are called on to work well beyond the forty-hour workweek, sometimes clocking eighty hours. Despite the widespread desirability of video game design jobs, dispatches from the hidden abode of production suggest that many are confined to regimented, repetitive tasks that some former employees have compared to factory work. An infamous anonymous posting on *Livejournal* in 2004 offered a glimpse at these practices. The anonymous author, "EA Spouse," complained about the mandatory unpaid overtime that was demanded of game designers at EA Games, a major sports video game developer. It was later revealed that the author of the post was the fiancée of an employee who was bringing a civil suit against the company. EA settled the suit out of court and reclassified two hundred employees so they were eligible for overtime, but left them ineligible for stock options. So, while the issue of labor conditions was addressed by EA, it was at the expense of the reward system that supposedly allowed video game workers to enjoy the financial success of the company.[24]

Given such conditions, the burnout rate can be high, but video game companies know that they can always replace workers with new hires who will be ecstatic to have landed their dream job. In this way, the video game industry treats labor in a manner similar to the sports industry, where labor is extracted from a continually refreshed stock of players. As many popular observers of sport have noted, sports organizations treat their athletic labor as eminently disposable, hence the frequent comparisons to a "meat market," where player labor is openly acknowledged and even celebrated as a "commodity" to be replaced by the next available player. In U.S. college sports, this exploitation is particularly stark, as the players are unpaid while participating at great physical risk. But even in the professional ranks, where financial rewards can be substantial, careers are on average very brief, and the physical toll, though often out of view, can be literally crippling.[25]

Organized attempts by athletes to improve working conditions also impact the financial bottom line for sports video game developers. As a result of partnerships with professional sports leagues around the world, EA Sports and related companies have a stake in the work stoppages that periodically interrupt play. In June 2011, for example, as a lockout threatened the upcoming NFL season, EA Sports president Peter Moore estimated that a canceled season would cost the developer seventy to eighty million dollars. Fortunately for the company, the lockout ended in late July, before the season began, though EA Sports pushed back the release date for *Madden NFL*. According to *Sports Business Journal*, the NFL agreed to a reduced licensing fee for the developer as a result of the stoppage.[26]

This relationship between labor issues in professional sports and the financial impact for video game developers is not always predictable or straightforward, however. For example, EA Sports' professional hockey simulation, *NHL 13*, enjoyed a substantial increase in game sales around its September 2012 launch, despite a lockout that delayed the season into 2013.[27] In addition to close ties with sports leagues, EA Sports also partners with the NCAA to produce simulations of American college football and basketball. EA Sports sponsors a preseason basketball tournament in Maui and in its video games reproduces the likenesses of NCAA athletes. Although the O'Bannon suit mentioned earlier is directed toward the NCAA, EA Sports is certainly a party of interest.

Clearly, the video game and sports industries are connected in many complex ways, and this collection begins to explore some of those complexities. The chapters contained within investigate the overlap by looking at a variety of issues, contexts, products, and modes of production. They do so by focusing attention through a critical lens and by asking how relations of power are articulated around and through the market and culture of sport-themed video games. The book is organized into two sections. The first considers some of the ways hegemonic versions of athletic masculinity are imagined in sports video games, but also considers how women are invited or incorporated into gameplay. The second section focuses attention on the various forms of simulation that can be found in sports video games, exploring the kinds of consumption, associations, and identification made available to the public in these games. At this point, we should explain these sections and the relationship of the various chapters in more detail.

GENDER PLAY

We have already discussed the way in which both sports and video games are implicated by gender roles, and in this section the authors further investigate the construction of gender in sports video games. The first chapters provide studies that explore how masculinity is produced through these new technologies and within a specific historical context, demanding new kinds of heroes, capable of transcending new challenges.

The section begins with Michael Z. Newman's historic account of "Jocktronics," in which he explains how the masculine gendering of video game play can be traced back to the sports video games that were introduced in the early years of the emerging video game industry. His chapter begins at the same point at which we began this introduction, the installation of *Pong* at Andy Capp's. Newman then considers the introduction and marketing of one of the first home gaming consoles, the Magnavox Odyssey. He also considers the way in which the popular press reported on video games and the commercials that were used to promote other early game consoles, including the Atari and Mattel's Intellivision. Consequently, Newman provides strong historical grounding for the other chapters that appear in this section.

Thomas P. Oates shows how masculinity transcends the marketing of console hardware and implicates one of the most popular video game software franchises by tracing the marketing techniques for EA Sports' *Madden NFL*. Reading ad campaigns, reality-program tie-ins, and promotional events, Oates identifies the *Madden* imaginary: a hypermasculine celebration that carefully navigates the subtly shifting ideological terrain of contemporary sport. Yet Oates extends the analysis of gender by engaging how the representation of masculinity intersects with the representation of race and speaks to white anxiety in relation to the role that black athletes play in professional football.

Gerald Voorhees's chapter extends the examination of gender by providing a deep analysis of what he describes as "the cultural practices of sportive gameplay." He begins with a theoretical overview of Michel Foucault's concept of governmentality, from which he derives a neoliberal version of subjectivity. As Voorhees observes, human action becomes legible, and intelligible, when specific performances are registered through the accumulation, coding, and comparison of data. In regard to sports and play, these data are readily apparent and ultimately used to determine who "wins." Because these data are mainly derived from the performances of men, they serve to measure masculinity. Voorhees's analysis begins with the older, more established theories about play in general, which he connects with more contemporary examples of digitalized and mediated play. Through this analysis, he is able to demonstrate that the gender overlap between sports and video games is more than a precious coincidence, but instead reflects a much more pervasive construction of masculine subjectivity.

Luke Howie and Perri Campbell take a departure from the study of video games to discuss National Basketball Association (NBA) fantasy league play. We include their chapter because it provides a strong connection between the gender play of sports video games and the larger sports-fan culture. After all, fantasy league sports often entail digitally mediated interaction (the league Howie and Campbell study is housed on Yahoo!) and thereby share an important similarity to sports simulation video games. Their analysis is more ethnographic in nature, and they include interviews with not only men who are active in NBA fantasy leagues, but also the wives and partners of fantasy league players. They conclude that these women provide a supportive role to their partners, and thereby have an important role in fantasy league play, a role overlooked by other studies of fantasy leagues.

Renee M. Powers and Robert Alan Brookey take the discussion of women and sports video games a step further in their analysis of the Nintendo Wii. They contextualize the Wii within the industrial conditions of Nintendo's "Blue Ocean Strategy." This approach was designed to market Nintendo's products to segments that fall outside of the traditional demographic of video games, specifically older adults and women. Powers and Brookey focus on Nintendo's efforts to market the Wii to women. Their analysis engages the actual design of the console and the software interface, as well as the marketing campaign and the *Wii Sports* and *Wii Fit* packages. The association of sports and fitness is so close that it has almost become a metonymy. This association has also been monetized, with both sports franchises and players branding and endorsing various fitness products. In fact, Nintendo's Wii console has been instrumental in allowing this sports and fitness association to be effectively monetized in the video game industry. Powers and Brookey argue that while opening the video game market to women may seem like a progressive move, the Nintendo Wii and its products reflect a view of women that is decidedly postfeminist and politically retrograde.

THE USES OF SIMULATION

The essays in the second half of the book explore the wider consequences of the pursuit of realism in sports video games. While other games construct imaginary worlds, dreamscapes, reimagined pasts, or unknown futures, sports game developers pride themselves on constructing the most convincing facsimile possible. The relentless pursuit of "authentic" simulations in these products works to craft sophisticated acquisitive fantasies, where material goods, brands, bodies, organizations, and even national identities are presented for consumption.

Steven Conway takes up the issue of simulation in an analysis of soccer stars as avatars, or, as he describes them, "celebrity avastars," in professional soccer simulation video games. Like Voorhees, Conway navigates the reader through several theoretical touchstones in order to comprehend the way sports celebrity is articulated (and rearticulated) in video games. He suggests that the deployment of these avastars within the practices of gameplay serves to highlight and amplify their masculine powers through their digitalized performances. Conway discovers in his analysis that "by transforming the celebrity into a hyperludic game piece primed for heroic acts, the developers propagate a form of cult worship synchronic with mass-media production."

Cory Hillman and Michael L. Butterworth examine the rhetorical means by which the goal of authentic simulation constructs a market-place for brands already circulating in mediated sport, while also advancing the video game producer's own institutional objectives. In the alternate reality of officially licensed sports video games, certain brands are highlighted, reinforcing their supposedly natural visibility at the center of what sportswriter Robert Lipsyte has called "Sportsworld," while exploiting commercial partnerships that lie out of view.[28] Andrew Baerg's chapter explores the EA Sports franchise *FIFA,* a soccer simulation that is one of the most profitable sports-based video game franchises. Baerg's analysis centers on the fleeting "border-crossing" practices offered by the game in a number of ways that work to construct varied and complex experiences of banal cosmopolitanism.

Meredith M. Bagley and Ian Summers consider the simulation offered by the EA Sports game *NCAA Football,* which selectively highlights certain elements of college football, while obscuring others. These selective inclusions and exclusions, they argue, promote the interests of the NCAA by fostering a familiar vision of the college sport through the experience of gameplay.

In the final chapter of the book, David J. Leonard, Sarah Ulrich-French, and Thomas G. Power consider how the Nintendo Wii console and its fitness- and sports-related offerings may (or may not) overcome socioeconomic barriers to fitness and exercise. Their chapter includes an extensive review of the literature on the Nintendo Wii, focusing on those studies that have attempted to assess the health benefits of the console and its games. They conclude that while the Wii may provide exercise opportunities to those often marginalized from health and fitness activities, those opportunities are not always realized.

The chapters collected here do more than add to the little-studied but crucial topic of sports video games. The scholarship in this book connects this growing and important sector of the industry to the broader cultural struggles in contemporary economic, cultural, and political life. In the decades since *Pong* debuted in the early 1970s, a profound set of shifts in economic policies, political initiatives, and cultural logics has generated a "neoliberal" condition, characterized by privatization, a shrinking welfare state, and changing attitudes about race, gender, and other aspects of identity. These shifts have worked to construct new models of "cultural citizenship": a new set of orientations, responsibilities, and obligations. In many ways, the connections between sport and video games can be understood in relation to a larger neoliberal project. We offer these chapters in order to begin identifying those connections and initiate a broader conversation about their relationship to contemporary social changes that have both positive and negative implications.

NOTES

1. Steven L. Kent, *The Ultimate History of Video Games* (New York: Three Rivers Press, 2001).

2. David Winter, "Magnavox Odyssey: The First Home Video Game Console," 2010, http://www.pong-story.com/odyssey.htm#P3.

3. Kent, *Ultimate History of Video Games*.

4. Donald Melanson, "A Brief History of Handheld Video Games," *Engadget,* March 3, 2006, http://www.engadget.com/2006/03/03/a-brief-history-of-handheld-video-games/.

5. Entertainment Software Association, "Essential Facts about the Computer and Video Game Industry: 2011 Sales, Demographic, and Usage Data," http://www.theesa.com/facts/pdfs/ESA_EF_2011.pdf. Although the video game industry generates revenue from sales of hardware (the game consoles) and software (the individual games), where the sport simulation games are concerned, revenue is generated in software sales, and these sales often occur across the various game console platforms.

6. See, for example, John Sherry, "The Effects of Violent Video Games on Aggression: A Meta-Analysis," *Human Communication Research* 27 (2001): 409–431; Michael Slater et al., "Violent Media Content

and Aggressiveness in Adolescents: A Downward Spiral Model," *Communication Research* 30 (2003): 713–736; Stacey Smith, Ken Lachlan, and Ron Tamborini, "Popular Video Games: Quantifying the Presentation of Violence and Its Context," *Journal of Broadcasting and Electronic Media* 47 (2003): 58–76.

7. Mark Wolf, *The Medium of the Video Game* (Austin: University of Texas Press, 2001); Diane Carr et al., *Computer Games: Text, Narrative, and Play* (Cambridge: Polity Press, 2006); Geoff King and Tanya Krzywinska, eds., *Screenplay: Cinema/ Video Game/Interfaces* (London: Wallflower Press, 2002); Justine Cassell and Henry Jenkins, "Chess for Girls? Feminism and Computer Games," in *From Barbie to Mortal Kombat: Gender and Computer Games,* edited by Justine Cassell and Henry Jenkins (Cambridge: MIT Press, 1998), 2–45; Yasmine Kafai et al., eds., *Beyond Barbie and Mortal Kombat: New Perspectives on Gender and Gaming* (Cambridge: MIT Press, 2011).

8. Edward Castronova, *Synthetic Worlds: The Business and Culture of Online Games* (Chicago: University of Chicago Press, 2005); T. L. Taylor, *Play between Worlds: Exploring Online Game Culture* (Cambridge: MIT Press, 2006); Ian Bogost, *Persuasive Games: The Expressive Power of Videogames* (Cambridge: MIT Press, 2007); Stephen Kline, Nick Dyer-Witheford, and Greg De Peuter, *Digital Play: The Interaction of Technology, Culture, and Marketing* (Montreal: McGill-Queen's University Press, 2003).

9. Robert Alan Brookey and Kristopher Cannon, "Sex Lives in Second Life," *Critical Studies in Media Communication* 26 (2009): 145–164; Lisa Nakamura, "Queer Female of Color: The Highest Difficulty Setting There Is? Gaming Rhetoric as Gender Capital," *Ada: A Journal of Gender, New Media, and Technology* 1 (2012), http:// adanewmedia.org/2012/11/issue1 -nakamura/.

10. Erica Rosenfeld Halverson and Richard Halverson, "Fantasy Baseball: The Case for Competitive Fandom," *Games and Culture* 3 (2008): 286–308; Mark Wolf and Bernard Perron, *The Video Game Theory Reader* (New York: Routledge, 2003); Mark Wolf and Bernard Perron, *The Video Game Theory Reader 2* (New York: Routledge, 2009); Wolf, *Medium of the Video Game,* 132.

11. There have been a handful of exceptions to this general trend. See, for example, Darcy C. Plymire, "Remediating Football for the Posthuman Future: Embodiment and Subjectivity in Sport Video Games," *Sociology of Sport Journal* 26 (2009): 17–30; Garry Crawford and Victoria K. Gosling, "More than a Game: Sports-Themed Video Games and Player Narratives," *Sociology of Sport Journal* 26 (2009): 50–66; Steven Craig Conway, "Starting at the 'Start': An Exploration of the Nondiagetic in Soccer Video Games," *Sociology of Sport Journal* 26 (2009): 67–88; David J. Leonard, "Performing Blackness: Sports, Video Games, Minstrelsy, and Becoming the Other in an Era of White Supremacy," in *Re: Skin,* edited by Mary Flanagan and Austin Booth (Cambridge: MIT Press, 2007), 321–339.

12. Aphra Kerr, *The Business and Culture of Digital Games: Gamework and Gameplay* (London: Sage, 2006).

13. Ibid.

14. Robert Alan Brookey, *Hollywood Gamers: Digital Convergence of the Film and Video Game Industries* (Bloomington: Indiana University Press, 2010); Alexis Blanchet, "A Statistical Analysis of the Adaptations of Films into Video Games," *INA Global: The Review of Creative Industries and Media,* July 12, 2011, http://www

.inaglobal.fr/en/video-games/article
/statistical-analysis-adaptation-films
-video-games.

15. Karen Raugust, *The Licensing Business Handbook* (New York: EPM Communications, 1995).

16. Tim Surette and Curt Feldman, "Big Deal: EA and NFL Ink Exclusive Licensing Agreement," *Gamespot,* December 13, 2004, http://www.gamespot.com//news/big-deal-ea-and-nfl-ink-exclusive-licensing-agreement-6114977.

17. EA Sports is being very vague about the details of their continued agreement with the NFL. Kahlil Thomas, "EA Sports NFL License: CEO Andrew Wilson Says That Company 'Has Agreements in Place' for Future Madden Games," *International Digital Times,* January 29, 2014, http://www.idigitaltimes.com/articles/21686/20140129/ea-sports-nfl-license-ceo-andrew-wilson.htm.

18. Chloe Albanesius, "EA Settles Football Class-Action Suit for $27 Million," *PC Mag.com,* July 23, 2012, http://www.pc-mag.com/article2/0,2817,2407524,00.asp; Matthew Coe, "Don't Expect a Change in the NFL License Status Quo," *Operation Sports,* March 27, 2013, http://www.operationsports.com/features/1621/dont-expect-a-change-in-the-nfl-license-status-quo/.

19. Actually, EA Games has a webpage devoted to mobile games, which feature cell phone platforms at the top of the page; there is even a special section for iPod and iPhone games. EA Sports licensed games are also featured, including several versions of *Madden NFL,* FIFA (Fédération Internationale de Football Association), NCAA, and even ESPN.

20. Michael Oriard, *Reading Football: How the Popular Press Created an American Spectacle* (Chapel Hill: University of North Carolina Press, 1993); Margaret Morse, "Sport on Television: Replay and Display," in *Television: Critical Concepts in Media and Cultural Studies,* edited by Toby Miller (London: Routledge, 1993), 383; David Theo Goldberg, "Call and Response: Sports Talk Radio and the Death of Democracy," *Journal of Sport and Social Issues* 22 (1998): 219; David Nylund, "When in Rome: Heterosexism, Homophobia, and Sports Talk Radio," *Journal of Sport and Social Issues* 28 (2004): 136–168.

21. The most recent report from the Electronic Software Association indicates that of the most frequent purchasers of video games, 48 percent were female. Electronic Software Association, "Essential Facts about the Computer and Video Game Industry: 2012 Sales, Demographic, and Usage Data," http://www.theesa.com/facts/pdfs/ESA_EF_2012.pdf.

22. Murray Sperber, "Why the NCAA Will Play On," *Inside Higher Ed,* February 18, 2013, http://www.insidehighered.com/views/2013/02/19/ncaa-flawed-politically-invulnerable-essay.

23. Kline, Dyer-Witheford, and De Peuter, *Digital Play.*

24. Tim Surette, "EA Settled OT Dispute, Disgruntled 'Spouse' Outed," *GameSpot,* April 26, 2006, http://www.gamespot.com/news/6148369.html; Robert Alan Brookey, *Hollywood Gamers: Digital Convergence in the Film and Video Game Industry* (Bloomington: Indiana University Press, 2010).

25. William Nack, "The Wrecking Yard," *Sports Illustrated,* May 7, 2001, http://sportsillustrated.cnn.com/si_online/news/2002/09/11/wrecking_yard/.

26. Stephany Nunneley, "EA Could Lose $80 Million in Revenue Should NFL Lockout Lead to a Cancelled Season," *VG247,* June 9, 2011, http://www.vg247.com/2011/06/09/ea-could-lose-80

-million-in-revenue-should-nfl-lockout
-lead-to-cancelled-season/; Daniel Kaplan, "NFL Gives EA a Break," *Sports
Business Journal,* February 14, 2011, http://
www.sportsbusinessdaily.com/Journal
/Issues/2011/02/20110214/Leagues-and
-Governing-Bodies/NFL-EA.aspx.

27. Brendan Sinclair, "NHL 13 Sales
Up Despite Lockout," *Games Industry*

International, September 20, 2012, http://
www.operationsports.com/forums/ea
-sports-nhl/579165-nhl-13-sales-up-year
-despite-nhl-lockout.html.

28. Robert Lipsyte, *Sportsworld: An
American Dreamland* (New York: Quad-
rangle / New York Times Book, 1975).

PART ONE

Gender Play

ONE

The Name of the Game Is Jocktronics: Sport and Masculinity in Early Video Games

Michael Z. Newman

ALTHOUGH IT MAY NEVER BE SETTLED WHICH VIDEO GAME deserves to be called the first, it's notable that two games based on racquet sports always come up in talk of the medium's origins. *Tennis for Two,* a demonstration using an analog computer and an oscilloscope at Brookhaven National Laboratory (1958), and *Pong,* the first hit coin-operated game from Atari (1972), are in some ways quite similar.[1] Both are competitions between two players given the ability to direct the movement of a ball, which bounces back and forth between them. Both are examples of sports games, a genre that would prove to be among the most enduring, enjoyable, and lucrative in the history of electronic play. And both can be placed within a tradition of masculine amusements adapted from professional athletics, which had already been popular in American society in penny arcades and around gaming tables for more than a half century when electronic games were new. We can regard *Pong* not just as an early and influential video game, but as part of a history of sports simulations and adaptations and as an electronic version of tavern and rec room amusements such as pool and Ping-Pong, from which it gets its name.

According to some historical accounts, the triumph of the *Pong* prototype at Andy Capp's tavern in Sunnyvale, California, launched a new medium in popular culture, marking the emergence of a new format of electronic amusement and a break from the past.[2] By offering an interactive experience of play controlling a small square of light on a video screen, *Pong* and many similar games in public and in the home did come across as novel and exciting, so much so that according to legend, the original game stopped working when it became overstuffed with coins. Games like *Pong* were a new use for the familiar cathode-ray tube T V set and a step forward in consumer electronics and high-tech leisure. To see *Pong* as a first, in retrospect, is to regard its difference from what came before it.

But the introduction of electronic games marked continuity as well as change. There is a tendency in popular media histories to see firsts as wholly original creative outbursts on the order of the Big Bang – at one moment there is nothing, and at the next we have the telephone or cinema or video games, as we would later understand these media. The reality is usually more messy and less teleological, as technologies and formats develop under the influence of available and familiar models. The fact that *Pong* was first offered to the public at a bar puts it into a context of coin-operated amusements such as jukeboxes, pinball games, and pool tables. As a racquet game it was instantly familiar, just as sports games based on baseball, football, horse racing, and boxing had been to thrill seekers at the penny arcade. In the same way that early cinema emerged out of nineteenth-century shows integrating imagery on screens, early video games drew upon forms of amusement and popular culture of their time.[3]

As video games emerged in the 1970s in the United States, many experiences of this new kind of electronic, mediated play involved some element of sports and relied on the player's familiarity with the most popular types of competitive play. Early games like *Pong* often had sporting themes. The first several years of home video game history are dominated by "ball-and-paddle" games, which expanded the basic racquet-sport concept to include versions of hockey, soccer, and handball and to add more sound and color.[4] The first home video game, the Magnavox Odyssey, included not only tennis but also hockey and football (as well as several other games). Other popular genres in these years included racing and shooting. With the introduction of programmable consoles such as Atari and Intellivision, which accepted game cartridges sold separately, many of the titles on offer were versions of professional team sports such as soccer, baseball, football, basketball, and hockey, and a number of auto racing games became popular. As competitive amusements calling on the skill and strategy of players using their bodies in pursuit of high achievement, video games themselves were quite sport-like independent of their content and were often discussed as a form of competition not so different from barroom sports such as darts and billiards or country club sports such as tennis. This distinguished them as a more active and masculine use of video technology than watching television, which by contrast was represented as passive. In some ways, video games drew on a heterogeneous array of influences for their content, from pro sports to science fiction as well as pinball and other coin-operated amusements. But despite this heterogeneity, competitive sports were of central importance to the development of their identity.

The affinity between sport and video games can help us understand the development of the cultural status of the emerging medium. We can use this correspondence in putting together a picture of the developing place of video games in popular imagination.[5] Video games developed in these early years into a form of boy culture, drawing on a tradition of masculine play and leisure-time amusement. Despite a tendency to see a new medium as a rupture from the past, as a revolutionary technological advance, video games can also be regarded as a way of playing some of the same sports and games as had been played in American society for decades, mediated through new instruments. The centrality of sports to early video games was one of the reasons the new medium was understood to be active, competitive, and masculine.

In this chapter I will consider three ways in which video games came into this identity during their early years, roughly 1972–1982. First, taking the example of the Odyssey, a game console released in 1972, I place video games into a history of masculinized sports simulations, games played both in public and in private. Second, looking at 1970s popular press discourses, I identify a recurring theme in these discussions of sports games representing a redefinition of television as an active, participatory medium, marking its transition from passive commercial broadcasting to active competition. And third, I analyze two TV advertising campaigns of the late 1970s and early 1980s for Atari and Intellivision. In these commercials, sports games figure prominently and cement the appeal of the new medium as a way of simulating professional athletics. In all three examples, video games are understood as a way of combining traditions of masculine amusement with electronic media technology.

ODYSSEY BETWEEN OLD AND NEW SPORTS SIMULATIONS

The Magnavox Odyssey has a special place in video game history, functioning as another first – first game for use with a consumer's TV set – and
also as a bridge between two kinds of play. Odyssey is the clearest example of the debt early games owe to the amusements of the pre–video
game years. Odyssey came on the market in the same year as *Pong,* and
the two have a common origin point in Ralph Baer's Brown Box, now in
the collection of the Smithsonian. Baer created this TV game prototype
for the electronics firm Sanders Associates in 1967 and 1968; Magnavox
later acquired the rights to market the technology.[6] But before it was
released it had been demonstrated publicly, and Nolan Bushnell and
Al Alcorn created *Pong* in imitation of Baer's invention, which they had
seen. Odyssey functions historically as a point of connection between
two technological eras, incorporating elements of both board or cabinet
games and video or electronic games. By contrast, *Pong* is self-contained
as an electronic game.

Bushnell and Alcorn had previously failed to arouse significant interest in an earlier coin-op cabinet game called *Computer Space,* modeled
after the mainframe game *Spacewar,* which Bushnell played as a student
in Utah during the 1960s. *Computer Space* was thought to have failed
because it was too challenging or confusing, and *Pong* was famously
simple: its only instruction was "avoid missing ball for high score." By
basing the game on the familiar concept of racquet sports, the player was
oriented instantly and understood the objective and operations. Odyssey and *Pong* both worked in more or less the same way, giving the player
operation of a paddle to use in directing the ball against an opponent's
paddle. Unlike *Pong,* however, Odyssey came packaged with an array
of other games. Odyssey *Tennis* was equivalent to *Pong* and required no
elaborate instructions or paraphernalia. The rest of the games one could
play on the Odyssey by inserting game cards into the console did require
these accouterments, sometimes requiring elaborate ensembles of paper
and plastic game components of the kind found in board game boxes.

In a number of ways, early video games picked up on traditions and
ideals of play that already existed and were well established in the 1970s,
particularly those associated with the suburban recreation room. Base-

ment sports such as Ping-Pong and pool and table games like cards and Monopoly were supposed to bring the middle-class family together in companionate leisure. Advertisements for early games including Odyssey and Atari often pictured family members together in the family or living room, united in electronic play, much as earlier ads for televisions and radios had pictured a family circle integrating all members of the household together in the pleasure of each other's company. Images of video games in department store catalogs of the 1970s placed the electronic amusements alongside other rec room products such as bumper pool and Ping-Pong tables. Sometimes the video games were pictured alongside other tabletop games like chess (or electronic chess). Imagery in magazines and catalogs pictured players in sociable settings, combining participants of mixed age or gender. Similar images were often seen in television commercials, including one Atari ad from the later 1970s in which a mother repeatedly rejects inquiries from babysitters, preferring to stay at home with her husband, son, and daughter playing video games.

At the same time as home play was represented integrating the family, throughout the postwar years some games popular for home play were addressed particularly to boys and men. These games often have professional sports themes. Numerous versions of baseball, football, basketball, horse racing, auto racing, boxing, and bowling were released in these years as board games. These might use the traditional board game materials of cards, dice, and tokens, but some were also similar to penny arcade cabinet games and integrated mechanical, electrical, and electronic components.[7] Penny arcade sports games go back at least to the 1880s, when automatic machines for shooting and horse racing were introduced. These games, like many similar coin-operated machines, proved popular with gamblers. Numerous penny- or nickel-in-the-slot sports games were fixtures in arcades beginning in the 1920s, when freestanding cabinet football games were particularly popular alongside boxing, racing, and other types of mechanical amusement.[8]

Compared with the physical skills demanded of penny arcade versions, some board games were considerably more cerebral and strategy oriented, such as APBA Baseball, a predecessor of Strat-o-Matic, Rotisserie, and fantasy sports leagues. APBA, first sold in 1951, claimed to be "scientific" and employed real Major League player statistics to allow for simulated games, series, and even seasons.[9] But sports games for the home could also be physically oriented, such as those using balls, levers, figures, buttons, lights, and other moving objects. Jim Prentice Electric Baseball (1953) combined buttons and lights with tokens moved by the players around a metal baseball diamond board. Tru Action Electric Baseball Game (1958) employed a magnetic ball and a spring-loaded bat, and play involved aluminum ballplayer tokens. ABC Monday Night Football, a board game released in the same year as Odyssey, boasted "computerized" play and employed a combination of cards, buttons, lights, and a plastic green football "arena" field. The board in these games represented a field (or track), and the rules and processes of the real sport were reproduced and simulated using paper, plastic, and metal objects, sometimes including electrical components. Whether focused more on reproducing the physical performance or intellectual strategy of professional athletics, games based on sports offered players a simulation of the real thing and a taste of the thrill of high-level athletic competition.

The Odyssey is a device very clearly looking both back and forth in time, integrating elements of rec room games of the 1950s and '60s with those of an increasingly electronic age. The original Odyssey console released in 1972 was primitive by comparison even to later ball-and-paddle consoles, and Ralph Baer evidently thought that the way it straddled the board game and TV game forms was a failing.[10] Historically, however, this ambiguity in form demonstrates how video games drew upon a context of pre-electronic games in establishing their identity, and it reveals continuity as well as disruption in the history of leisure and amusement.

The electronic component of gameplay in Odyssey games was often rather slight, and the games required additional nonelectronic materials to be played at all. In most of the original Odyssey games, each of two players could control a square of light by moving it vertically and horizontally, and sometimes the player could put "English" on a ball that bounced between the rectangles, causing it to curve rather than travel straight. Odyssey had neither sound nor color. In some games, the image on screen could be made to move somewhat at random to come to a stop at a point on the display, an effect similar to spinning a wheel or rolling dice. To complete its representation of game spaces, the Odyssey came packaged with translucent plastic overlay sheets that adhered to the glass of the CRT screen by its static electricity charge. Odyssey *Tennis* had a green court overlay, *Hockey* was a white rink, *Football* a green field, *Roulette* a red and black wheel, and so on. These overlays remediated the tabletop sports games like Electric Baseball, adapting them for play with a television set. Odyssey also came packaged with a variety of paraphernalia, including game chips, cards, dice, paper money, and game boards. In *Football,* the electronics functioned as one component of a wider ensemble of devices and materials, including a cardboard football field similar to the fields in sports-themed board games. *Football* players would sit across from each other with this field between them on the table. Playing Odyssey *Football* also required a paper scoreboard, a roll of frosted tape, a football token, a yardage marker, and six separate decks of cards for different kinds of plays, including passing, running, and kicking off. The manual spent six pages describing the process of gameplay, and players would have needed to keep it open while playing at least at first. Odyssey *Football* bears a strong resemblance to the football games for home play sold over several decades before electronic TV games came along.

As its game titles suggest, Odyssey, like many early consoles, was aimed at families. *Haunted House, Cat and Mouse,* and *Simon Says* are juvenile in their cartoonish representations and in their cultural associations. *States* and *Analogic* were meant to be educational. *Shooting Gallery,* a popular title using a rifle controller, was similar to the basement Daisy BB Gun ranges marketed to fathers and sons in 1960s issues of *Life.* Sports games might appeal to boys seeking indoor diversion on rainy days, perhaps in the company of parents, siblings, and friends. That most of these games were designed for two players rather than one indicates the sociable intentions of the producers and advertisers. Like later video game consoles and titles, the original Odyssey appealed to the suburban, middle-class family as a means of bringing them together. But in its continuation of traditions of masculine sports-themed play, Odyssey was also a way of further mediating and simulating the competitive amusements of boy culture, integrating the television set into the ensemble of technologies used for sporting leisure. From the start, home video games were helping children to fashion "virtual play spaces" where they would continue the modes of gendered fantasy and interaction that date at least to the nineteenth century. For boy culture these would include competitive feats of physical achievement modeled on adult roles (professional athletics in this case) and performed for the recognition of peers.[11] In maintaining this tradition, Odyssey was not so different from the sports games that preceded and influenced it.

VIDEO GAMES, SPORTS, AND TELEVISION: MASCULINIZING A PASSIVE MEDIUM

Early video games had not yet established a well-defined identity as a medium, and efforts to understand them often struggled to relate video games to more familiar artifacts and experiences. New media frequently remediate old media, incorporating their contents and formats.[12] Video games built on older forms of play from arcades and family rooms; they also remediated television, their technological component most familiar to early users. Their very name incorporates the word *video,* often used as a synonym for television itself. Other names used in the 1970s included tele-games and TV games. The remediation of television by electronic games was understood not merely as a new use for the TV set, but as its transformation into something radically improved.

Early video games were often discussed, as new media technologies often are, in techno-utopian terms. They were thought to promise a solution to some of the problems with television. American television in the 1970s was occasionally admired for some of its evening programming such as the socially "relevant" hits *All in the Family* and *The Mary Tyler Moore Show,* aimed at elite audiences. The medium's reputation was nevertheless indisputably low. These were years when books appeared with titles such as *Four Arguments for the Elimination of Television* and *The Plug-in Drug.* Following Federal Communications Commission chairman Newton Minow's "Vast Wasteland" speech in 1961, the cultural status of television was more that of a social problem than an art form. It was widely seen as a debased commercial mass medium devoted to profit and little else, a lowest-common-denominator "chewing gum for the eyes" to serve up audiences to advertisers, and a way for children, housewives, and shut-ins to pass their time. Rather than engaging audiences intellectually, in the manner of literature and other legitimate arts, television was believed to lull audiences into a kind of complacent, distracted passivity.[13]

In remediating television, early video games were often seen to be transforming and redeeming TV and marking its transition out of passivity and into activity. For many, this put video games in the same place culturally as other media innovations of the time, such as video art, cable television, and home videotape machines, all of which were seen to be making similar interventions into television as a mass medium.[14] Characterizing media technology and culture in these value-laden terms has strong undertones of gendered distinction, as the feminized passive medium was to be improved by being made more active, that is, by being masculinized.[15] The sports element of early games was central to this emerging discourse of video games improving TV by making it active. Sports was both literally the content of early games, but also a figure for understanding what the experience of games would be: rather than merely watching passively, the audience for TV would be transformed by games into athletes in active competition.

Frequently, when journalistic discussions tried to explain this new form of electronic amusement to readers, the sports themes of many early games helped make a distinction between television as a broadcasting medium and television as one component of the new games. In a *New York Times* story about the release of the Odyssey in May 1972, for example, the president of Magnavox (a longtime manufacturer of television sets) claims his company's product "is an educational and entertainment tool that transfers television from passive to an active medium." A *New York Times Magazine* feature two years later included quite similar language: "Odyssey costs $99.95 and is simple enough for the average consumer to attach to the back of his TV set, transforming the otherwise passive box in his living room into a game screen with two 'player' blips and a 'ball' blip moving across it." The article first describing TV games for readers of *Time* in 1972 paints the following contrast: "The average American spends six hours a day gazing passively at television. Soon he will have an opportunity to play a more active role in what appears on the screen of his set. Last week the Magnavox Co. demonstrated a device that will give set owners a chance to engage in electronic table tennis, hockey, target shooting and other competitive games on their TV screens."[16] Although many of the Odyssey games included with the console had no sports theme at all – some were educational, such as *States,* and some were typical children's games, such as *Simon Says* – the emphasis here was on the most competitive and familiar sports games that might appeal to the male breadwinner making major purchases for the middle-class suburban family.

Publications aimed at male electronics hobbyists were especially invested in this kind of remediation of television. *Radio-Electronics* placed games into a standard before-and-after scenario, representing a transformation of television from meager entertainment to a more elaborate fantasy world of immersive sporting simulation: "For a considerable number of years, we sat in front of our TV sets and let them entertain us with moving pictures on that little screen. . . . Yet there is a new kind of entertainment on that home TV screen – it's a Ping-Pong game, a soccer field, a shooting gallery and others and you, who until now have been a passive viewer get to control the action."[17]

In all of these accounts of the new technology, we find not merely that the television is used in a new way, but that the user's experience is understood to have a value diametrically opposed to the typical television audience's. Another hobbyist publication, *Mechanix Illustrated,* captures this totally different idea of the television audience, a fresh identity in contrast to the lazy spectator of the past: "TV screens used to be just for watching whatever the network or the local station felt like putting on the air. Now the home set has become the center of family sportsmanship. . . . It appears that the bouncing blip could change the habits of American TV gluttons. It will surely get them more involved."[18] By expressing or implying a contempt for television's passive, implicitly feminized audience, these discourses articulate video games with a number of related conceptual terms, from the more specific, such as sports and competitive games, to the more abstract, such as active, engaged, involved. The viewer is in control.

A 1976 trend piece on the video games craze in *Time* drew upon a familiar trope of writing about video games at this time, treating the new medium not so much as its own thing and more as a thing to do with a TV set. The headline is "TV's New Superhit: Jocktronics." The idea that a video game is a hit T V show confuses games, which you play, with shows, which you watch. And by calling this T V program "Jocktronics," the article makes clear that sports would be central to the remediation of television by gaming. The article offers this description of video game play: "Television's new superhit is Yourself, the Athlete (or Racing Driver, or Op Artist, or Blackjack High Roller). The name of the game – which is provided by a wide and wildly competitive assortment of electronic contests that can be simply hooked into any T V set – is Jocktronics."[19] By becoming the protagonist of the program, the viewer becomes an active agent, racing down a track or calling "Hit" at the card table. Television, so long understood to be a source of inane entertainment and diversion, was newly able to include the viewer not as a recipient of programming but as a part of the action. In presenting the confusion between sports play and television spectatorship, writers and illustrators would call upon familiar logics of television as passive and sports as active, suggesting that making T V a technology for play means renewal through greater activity. Jocktronics in particular masculinizes television, as the image of the jock is ever an essentially male one. But more generally by rejecting passivity, the discourse of video games as a form of sporting competition on television worked to distance the older medium from its traditional identity.

An especially important term in these discussions is *participation,* understood as a virtue then and now. We might think today that these ideas about media and play call to mind interactive media, but that term was not used to describe video games in the 1970s, and it would be anachronistic to apply that concept here; *interactive* in this usage comes from the field of computers, which entered Americans' everyday lives – at least as users – several years later than video games. *To interact* means that the machine communicates with you by giving output in instant response to your input, but *to participate* puts all of the agency onto the human user, whose involvement is a notable, fresh contrast against a previous lack of action. To call video games participatory and to say that their players were participants through technology associated the games with other practices and discourses of the 1970s, such as the craze for two-way C B radio and the rising popularity of physical fitness leisure such as jogging. To participate is to be involved and to contribute, rather than merely to receive. One of the main criticisms of television broadcasting was that it had been a one-way medium, and efforts to transform it typically involved ideas that would point toward audiences giving feedback rather than merely receiving a signal.[20] When an article in *Sales and Marketing Management* in 1976 promised, "With the video games, it's strictly a participation sport," it was tapping into a wider discourse of media technology, in which participation was a value deemed to be missing from the institutions and practices of mass culture. Atari's Nolan Bushnell told a magazine interviewer in 1974 that he wanted to make amusements be "not just spectator-oriented but participatory." Reporting on the popularity of electronic games as Christmas presents in 1975, the *New York Times* quoted a retail analyst asserting, "We may be leaving the spectator era for the participation era."[21] The analyst quoted in the *Times* claimed that thanks to participation replacing spectatorship, that year he was even planning to skip watching the Super Bowl!

The popular press can be breathless in its techno-utopianism, but it often also signals uncertainty and anxiety about new technology. Sometimes, popular press representations referenced a sense of confusion around the status of the new games. Were they really active forms of play like sports or passive forms of mass culture like T v? A *Newsweek* story on video games published in 1972 begins with a description meant to trigger this category confusion and probe the status of this new use for T v technology: "'Let's play football,' one kid suggests. 'Tennis!' insists the other. They compromise on hockey. But instead of gearing up and heading for the front door, they troop into the living room and flip on the television set." This description is meant to defamiliarize sports and put the new games into a recognizable scenario, but it is also a bit unsettling in its replacement of indoors for outdoors and electronics gear for fields and courts. In a similar vein, a cartoon by Ed Arno illustrating a 1978 article on video games in *Changing Times,* a mass circulation magazine, pokes fun at the idea of these games as a more active use of the T v set. Seven years into the history of home video games, perhaps the idea of transformation had become familiar. Arno pictures an adult couple dressed for tennis, wearing visors on their heads and with racquets by their sides, sitting on a sofa in front of a T v set playing *Pong.* The mismatch between their sports apparel and equipment and their activity speaks to the status of video games as a technology caught between authentic athletic competition and simulating sports through electronics.[22] Despite efforts to paint video games as participatory and active, their association with television would continue to influence their cultural status. But sports were central to their emerging identity, and the positive values often associated with video games in their early years were a product of the high regard for active participation and the usefulness of sports as an example of highly valued recreation.

INTELLIVISION, ATARI, AND BOY CULTURE FANTASY

By the beginning of the 1980s, video games had become familiar objects in public places, not just arcades but also bars, truck stops, bus depots, and any place where coin-operated amusements might be found. They were also a booming home entertainment business. Atari was acquired by Warner Communications in 1976 and quickly began to spend huge sums on advertising and promotion, particularly television commercials. Atari was the leader by far in this field, spending $18 million in 1981 to its nearest competitor's $5.5 million.[23] That competitor, the toymaker Mattel, had significant success in the later 1970s with its handheld electronic Football game, with its long rectangular field, red LEDs, and plastic buttons. This was at the same time that handheld electronics, from games such as Simon and Computer Perfection to educational toys such as Speak & Spell, were selling so briskly their manufacturers could not keep up with microchip supplies.[24] Mattel entered the programmable game market in 1979 and positioned itself as the most threatening rival to Atari in a lively field. While many other companies, including RCA, Fairchild, Bally, Magnavox, and Coleco, got into the programmable video game console business, it was two rival products, Atari's Video Computer System (later called 2600) and Mattel's Intellivision, that had the highest profile and most often represented home video games in popular imagination. The identities of both consoles were very much shaped by their use for playing sports games. Atari's controls and graphics were simple and even primitive by comparison to Intellivision's, which had not only a directional control but also a number pad and buttons on the sides. Atari had certain advantages, though: it boasted more titles (a selling point in ads), more hit arcade game ports (*Space Invaders, Asteroids, Missile Command, Pac-Man*), and a stronger video game brand identity. Intellivision's advertisements prompted comparisons to Atari, while Atari's power position meant they might ignore the competition.

These two consoles were fairly typical of the period in terms of their sports games offerings. The first programmable console for purchase in American retail stores was Fairchild's Channel F, which appeared in 1976 with *Hockey* and *Tennis* built in. Other games available for Channel F as cartridges included *Shooting Gallery, Drag Strip, Baseball, Bowling,* and *Pro Football.* Consoles of similar vintage had similar titles. Baseball video games, to choose one popular sport, could be played on numerous programmable consoles of the 1970s and '80s, including the APF MP1000, Arcadia 2001, Bally Astrocade, Odyssey, and RCA Studio II, in addition to Atari and Intellivision.[25] Atari's baseball game *Home Run* was released earlier, in 1978, and Intellivision appealed on the basis of its *Major League Baseball,* released in 1980, boasting superior graphics and gameplay. Atari's game allowed for a variety of pitches, including fastballs, curveballs, screwballs, and changeups, but players often found many other aspects of the game wanting, particularly the defensive aspects of the game and the absence of certain key elements of play such as fly balls and throwing runners out at a base. Intellivision's baseball game improved on this by including base stealing and bunting, among other details. Using synthesized speech, an umpire's voice would be heard calling players out. More than a million cartridges sold, making *Major League Baseball* Mattel's top seller.

As in this comparison between two baseball simulations, one of the criteria for evaluating and appreciating sports video games in these formative years was their ability to represent the details of a professional spectator sport and to render the experience with a sense of realism. The rhetoric of authenticity is everywhere in discussions of these sports games, positioning the new medium as a mimetic technology for the faithful reproduction of experience electronically. One way in which Intellivision sought an edge in this regard was by licensing the identities of the major league sports organizations, so that in comparison to cartridges merely called *Golf* or *Football,* Mattel offered official-sounding versions complete with familiar logos on the cartridge and in catalog art.

Atari's sports games included *Basketball, Bowling, Boxing, Championship Soccer, Football, Golf, Ice Hockey, Racquetball,* and *Tennis.* Intellivision's were *Major League Baseball, NFL Football, NHL Hockey, NBA Basketball, NASL Soccer, PBA Bowling, PGA Golf, U.S. Ski Team Skiing,* as well as variations on these such as *World Series Major League Baseball* and *World Cup Soccer.* Intellivision's appeal was as the version closer than rival products to the game as played professionally. Both companies, however, were eager to associate their games with the highest levels of competition.

Atari's print and TV campaigns in the later 1970s had a number of related agendas. Some TV spots and display ads marketed the game console, while others hawked the game titles for sale separately. One particularly relevant campaign for this discussion had the tagline "Don't watch television tonight. Play it!" In both magazine and television advertisements, this promotion aligned the transformation of electronic media from passive to active with another change, from watching television to playing sports on television. This drew upon the rhetoric of participation familiar from the initial considerations of video games as a brand-new technology.

In the Atari TV spots, video games would invite players to act out fantasies of masculine empowerment as they compete at virtual sports against the most famous seasoned professionals they have watched on television. In appealing to children, these scenarios sold a pair of competitive, if comical, fantasies: playing in the big leagues and besting your heroes. "Don't watch television tonight" functioned as an entreaty to the masculine sports fan to enter the era of participation through electronic play, while leaving behind the debased mass culture of broadcast TV.

Commercials for this TV campaign featured the most famous male athletes whose skills on the field would comically fail to translate into video games. Kareem Abdul-Jabbar, star center of the Los Angeles Lakers, stands facing the camera under a basketball net, hands raised in defense. He says incredulously, "You gonna slam dunk me, Atari?" After picturing the gameplay in Atari's *Basketball,* Kareem sits, dejectedly cradling his head in his hand, while a little boy gloats over his victory. A baseball hero of the Cincinnati Reds stands at the plate with bat raised. Looking in the camera's direction he says, "Okay, Atari, let's see your best pitch!" Cut to the umpire: "You're out, Rose!" These humorous scenarios reinforced the activity of video game play by association with athletics and also empowered youthful players to defeat the pros in video simulations of real-life contests. The Brazilian superstar of the New York Cosmos confesses, "I quit soccer to play Atari," but in the next shot a little girl, perhaps his daughter, waves a finger at him, delivering her line in a sing-song voice, "You need more practice, Pelé!" The consumer is invited to develop a skill by playing games, just as one would with a serious sport like soccer or baseball. At the end of the thirty-second spot, a voice-over reminds consumers that the action in Atari would be "on your own T V set." The tagline is spoken: "Don't watch television tonight. Play it!" In such messages, the public image of video games was of something you do on your own T V set, but also of a means of making television do new and exciting and particularly active and competitive things. To "play T V" is to redress the failure of mass media and substitute a new sporting participation, with its promising dimensions of purposive engagement and user control, for the old experience of television as a one-way medium.

Intellivision's T V campaigns addressed similar concerns and fantasies. In every possible way, however, Intellivision sought distinction from its more powerful competitor. Rather than hiring the most famous pro sports heroes of the day, Intellivision cultivated a sophisticated and adult brand identity by choosing as a spokesman George Plimpton, the popular writer whose transatlantic accent announced his high social position. Plimpton had some sports bona fides, having written a first-person account of playing for the Detroit N F L team, *Paper Lion,* as well as essays on golfing and boxing and playing hockey and baseball against pro athletes (he had also performed with the New York Philharmonic and the Ringling Bros. Circus). The name Intellivision was a portmanteau of *intelligent* and *television,* and its tagline and slogan printed on product packaging was "This is intelligent television."[26] This promoted playing video games as something better than ordinary uses of T V, part of the effort at transformation of the feminized passive medium. The suggestion in the name Intellivision was that previously, T V had been an "idiot box," brainless mass culture. The premise of the T V ad campaign was to compare Atari and Intellivision products and show discerning consumers the obvious advantages of Mattel's alternative, especially in visible qualities of graphics and in the extent of the simulation's accuracy and detail. Some of these ads compared the most popular genre of arcade games of the late 1970s, which helped sell consoles: science-fiction space battles. But Mattel's television campaign also highlighted quite centrally a comparison of Atari's baseball cartridge and Intellivision's *Major League Baseball*.

Two commercials in particular, both of which appeared on TV around 1980, aimed to establish and reinforce sports games as key desirable elements of the home console. Both of them made the graphical display on the television screen into a main selling point, showing the clear contrast between Atari's sports simulation and Intellivision's. The rhetoric of these ads appeals to the authenticity and fidelity of representation. The intelligent consumer is naturally expected to prefer the choice that is closest to the real thing. The better game is the one that most faithfully re-creates the sport of baseball as played in the major leagues. This is hardly a new discourse, as the same logic of realism informed earlier games such as APBA and Strat-o-Matic, which were distinguished from rivals by statistical and scientific claims to fidelity. An investment in this quality was also familiar from the discourses of audio reproduction and home stereo electronics.

Both Intellivision TV spots feature comparisons of baseball games, while one also includes a comparison of the two brands' football titles. In both instances, the side-by-side, back-and-forth form recalls the classic comparative advertising strategy familiar, for instance, from cleaning products campaigns. In this instance, however, the point is not to show the superiority of one product over another in efficiency or power, but to illustrate the ability of an electronic simulation to re-create a professional sporting event as one would watch it on television or play it as a competitor. In the more simple and direct commercial, Plimpton faces the camera dressed in a jacket and tie, smiling at the audience. His class and intellect are the basis for Mattel's appeal to authority, inviting the audience's trust. "Here's an easy question for you," he begins. "Which of these things is closest to the real thing? A. Intellivision *Major League Baseball*. B. Atari baseball. Here they are again close up. A. Intellivision. B. Atari." The tone here is generously professorial, even Socratic; the question is so readily answered. It concludes, "If you thought A. Intellivision, you're absolutely correct. You see, I told you this question was easy." All the while, the image alternates shots of the games, in which the primitive qualities of the representation of Atari's version are contrasted against the greater detail in Intellivision's.

Another spot shows Plimpton sitting in a wing chair in what initially appears to be a den or club, though as the camera moves we see that it is more like a family room. He is in a jacket but with his shirt open at the neck, a well-off man at leisure. A graphic identifies him as a "famous author & gamesman." Children beside him are playing games on a pair of television sets, one Atari and the other Intellivision. "I'll try almost anything," he begins, calling on the audience's memory of his journalistic exploits. "So when Mattel Electronics asked me to compare their television games with Atari, I gave it a try. I compared Atari baseball with Intellivision and found Intellivision played much more like real baseball. Then I compared Atari football with Intellivision. Again Intellivision played more like the real game. In my opinion if you try them both, there's only one conclusion you can come to. Intellivision from Mattel Electronics." In the case of the football comparison, we are shown two representations of the game that are substantially different. Atari's version shows the field oriented vertically, with end zones at top and bottom and no marked yard lines. Intellivision's field is oriented horizontally, and numbered yard lines and player figures add more detail. The Mattel image looks a bit more like the image of televised football, though like Atari's it is rather abstract, particularly in comparison with the *Madden* game imagery of several decades later. The Mattel versions also differ in sound effects, offering synthesized bits of noise mimicking a ball sailing through the air and crowds cheering in comparison to Atari's generic beeps and blips.

Many early video game advertisements, for various types of games, addressed male players in particular. While the term *gamer* was not yet used as it would be in subsequent years, some of the gendered associations of that classification were falling into place.[27] The masculine gendering of video games had many causes. It is, moreover, an ongoing and complex dynamic that includes technologies, representations, and social practices and ways of understanding and that overlaps with other vectors of identity.[28] In the early period of console games when Intellivision and

Atari competed for the consumer's attention and money, the presumed masculinity of video games extended not only to typically male interests such as team sports, but also to the practices and values of boy culture that were not specific to sports per se, but would include any kind of competitive play. The idea that through new technology, one could play at a high level against great competitors, and that this experience would be like real sports, appealed to boy culture fantasies of physical achievement and mastery, recognition for these accomplishments, and role-playing adult identities.[29] Both the Atari and the Intellivision campaigns made video games into surrogates for athletic exploits outside of the rec rooms and basements where these amusements would be found, tapping into a long-standing practice of American boys (and boys in many other cultures) pretending to be their heroes on the playing fields and courts. Even as the newest high-tech gadgets, home video games were still selling old-fashioned fantasies of masculine self-actualization and achievement. The realism promised in ads for sports games was one important way in which the new medium developed as a form of boy culture.

CONCLUSION: SPORTS AND THE CULTURAL STATUS OF A NEW MEDIUM

Between the time when video games were introduced to the public in the early 1970s and the moment by which they had become a pop culture craze in the early 1980s, making the cover of *Time*, an identity was developing for the new medium.[30] A medium's identity, its cultural status and place in popular imagination, is a product of many factors and influences. It can be shifting and contradictory. Technologies are not destined to have particular uses or meanings, but arise to meet social needs and develop according to these, often in ways unforeseen by those who invent and produce them. In the early periods of any medium's emergence, we often find a certain flexibility in its identity. Video games were positioned through many competing and complementary discourses, including advertising and promotion, popular press coverage, product packaging, and the games themselves. They emerged into particular social contexts of amusement and leisure. As electronics, they had the cachet of the latest high-tech gadgetry. When they were new, it was not predestined that they would become so closely associated with youth and masculinity. Their first players were adults. Computer games emerged in institutional settings, in places like MIT and Stanford, and *Pong* had its debut in a bar. Early game advertising appealed to adults more than children and offered video games as family amusement to bring together players of various ages, both male and female.

Sports games were one important part of the story of video games' development toward the cultural status they would arrive at. The centrality of sports to the genres of early games, and to the conception of the video game console as a way of remaking the TV into a participatory cultural form, shaped the medium's identity. Early games remediated types of amusement and recreation that were already well established as masculine, such as simulation of professional sporting competition. They also made over the television set to integrate it into this tradition of gendered play. In doing so they renewed its identity, and it was reconceived as a playing field, a fresh conception for a familiar appliance.

This was part of a wider phenomenon in the 1970s and 1980s of the mediation of leisure and of the indoorsification of American childhood in particular and American life more generally, as suburban sprawl and fears about unsupervised children transformed the geography of American childhood.[31] Video games might have seemed like something new, and as technology they introduced previously unknown experiences of electronically mediated recreation. They also were a novel way of continuing established practices and the values and ideals associated with them. Sports video games, from *Pong* to *Major League Baseball* and beyond, presented new ways of doing old things. Jocktronics was not the only way of thinking about video games, but even if the name didn't stick, the connotations it expressed of high-tech, masculine, competitive amusement were powerful forces in shaping the medium as a cultural form, enduring long after Atari's and Intellivision's 8-bit sports cartridges had passed from the scene.

NOTES

1. On *Tennis for Two*, see Van Burnham, *Supercade: A Visual History of the* *Videogame Age, 1971–1984* (Cambridge: MIT Press, 2003), 28; and *When Games*

Went Click: The Story of Tennis for Two (2013), a documentary video directed by Vlad Yudin. On *Pong,* see Henry Lowood, "Video Games in Computer Space: A Complex History of *Pong,*" *IEEE Annals of the History of Computing* (July–September 2009): 5–19.

2. Two popular accounts are Roberto Dillon, *The Golden Age of Video Games: The Birth of a Multi-million-Dollar Industry* (Boca Raton, FL: A. K. Peters / CRC Press, 2011), 14–20; and Tristan Donovan, *Replay: The History of Video Games* (East Sussex, UK: Yellow Ant, 2010), 15–27.

3. Charles Musser, *The Emergence of Cinema: The American Screen to 1907* (Berkeley: University of California Press, 1994).

4. Leonard Herman, "Ball-and-Paddle Consoles," in *Before the Crash: Early Video Game History,* edited by Mark J. P. Wolf (Detroit: Wayne State University Press, 2012), 53–59.

5. My usage of "popular imagination" follows William Boddy, *New Media and Popular Imagination: Launching Radio, Television, and Digital Media in the United States* (Oxford: Oxford University Press, 2004).

6. For more on the Brown Box and Odyssey, see Ralph H. Baer, *Videogames: In the Beginning* (Springfield, NJ: Rolenta Press, 2005).

7. Robert Cantwell, "The Fun Machines," *Sports Illustrated,* July 4, 1977, 24–29.

8. Nic Costa, *Automatic Pleasures: The History of the Coin Machine* (London: Kevin Frances, 1988), 17, 137, 140–143.

9. Sandy Treadwell, "Dice Ball Keeps the Mind Fit," *Sports Illustrated,* November 17, 1969, 107–109.

10. Donovan, *Replay,* 22.

11. Henry Jenkins, "Complete Freedom of Movement: Video Games as Gendered Play Spaces," in *The Game Design Reader: A Rules of Play Anthology,* edited by Katie Salen and Eric Zimmerman (Cambridge: MIT Press, 2006), 330–363. "Virtual play spaces" comes from Jenkins.

12. Jay David Bolter and Richard Grusin, *Remediation: Understanding New Media* (Cambridge: MIT Press, 1999).

13. On the history of television's cultural status, see Michael Z. Newman and Elana Levine, *Legitimating Television: Media Convergence and Cultural Status* (New York: Routledge, 2012), 14–37.

14. Michael Z. Newman, *Video Revolutions: On the History of a Medium* (New York: Columbia University Press, 2014), 17–44. On discourses of cable television, see Thomas Streeter, "Blue Skies and Strange Bedfellows: The Discourse of Cable Television," in *The Revolution Wasn't Televised: Sixties Television and Social Conflict,* edited by Lynn Spigel and Michael Curtin (New York: Routledge, 1997), 221–242. On the relation between video games and video art, see Jason Wilson, "'Participation TV': Videogame Archaeology and New Media Art," in *The Pleasures of Computer Gaming: Essays on Cultural History, Theory, and Aesthetics,* edited by Melanie Swalwell and Jason Wilson (Jefferson, NC: McFarland, 2008), 94–117.

15. On the feminization of broadcasting and television in particular, see Boddy, *New Media and Popular Imagination;* Newman and Levine, *Legitimating Television;* and Lynn Spigel, *Make Room for TV: Television and the Family Ideal in Postwar America* (Chicago: University of Chicago Press, 1992).

16. "Magnavox Unveils TV Game Simulator," *New York Times,* May 11, 1972; Peter Ross Range, "The Space-Age Pinball Machine," *New York Times Magazine,* September 15, 1974; "Modern Living: Screen Games," *Time,* May 22, 1972.

17. Larry Steckler, "TV Games at Home," *Radio Electronics,* December 1975, 29.

18. Dick Pietschmann, "The New Fun World of Video Games," *Mechanix Illustrated,* January 1975, 36.

19. "TV's New Superhit: Jocktronics," *Time,* December 13, 1976.

20. Hans Magnus Enzesberger, *The Consciousness Industry: On Literature, Politics, and the Media* (New York: Seabury Press, 1974), 97.

21. "The Big Winners in Consumer Sales," *Sales and Marketing Management,* August 9, 1976, 23–25; Nolan Bushnell quoted in Donovan, *Replay,* 36; William D. Smith, "Electronic Games Bringing a Different Way to Relax," *New York Times,* December 25, 1975.

22. "Instant Replay," *Newsweek,* October 30, 1972, 75; "The Fancy New Video Games," *Changing Times,* November 1978, 43.

23. "Home Video Game Warfare Erupts on Television," *Broadcasting,* March 1, 1982, 64.

24. "Why Electronic Games Will Be Hard to Find," *Business Week,* November 17, 1979, 52.

25. A useful source for information about early video game cartridges is Brett Weiss, *Classic Home Video Games, 1972–1984: A Complete Reference Guide* (Jefferson, NC: McFarland, 2012).

26. Scott Sherman, "In His League: Being George Plimpton," *Nation,* February 2, 2009. On Intellivision's name and its effort to give television an "extreme makeover," see Sheila E. Murphy, *How Television Invented New Media* (New Brunswick, NJ: Rutgers University Press, 2011), 50.

27. Adrienne Shaw, "Do You Identify as a Gamer? Gender, Race, Sexuality, and Gamer Identity," *New Media and Society* 14 (2012): 25–41.

28. Michael Z. Newman and John Vanderhoef, "Masculinity," in *The Routledge Companion to Video Game Studies,* edited by Mark J. P. Wolf and Bernard Perron (New York: Routledge, 2014), 380–387.

29. Jenkins, "Complete Freedom of Movement."

30. "Gronk! Flash! Zap! Video Games Are Blitzing the World," *Time,* January 18, 1982.

31. Steven Mintz, *Huck's Raft: A History of American Childhood* (Cambridge, MA: Belknap Press of Harvard University Press, 2006), 347.

TWO

Madden Men
Masculinity, Race, and the Marketing
of a Video Game Franchise

Thomas P. Oates

IN AUGUST 2012, AS THE RELEASE OF EA SPORTS' *MADDEN NFL* 13 video game approached, a months-long marketing blitz peaked with a series of advertisements featuring actor Paul Rudd and Baltimore Ravens linebacker Ray Lewis. In the campaign, the two are presented as close, lifelong friends, whose bond is cemented by periodic *Madden NFL* marathons. The ads are clearly presented with tongue firmly in cheek. The friendship between Rudd and Lewis is offered as a whimsical premise. Rudd is a recognizable film and television actor, best known for roles playing middle-class white professionals. While appearing to be reasonably fit, he would never be mistaken for an NFL player, and though his movies are frequently about masculine themes (see, for example, *I Love You, Man; The 40-Year Old Virgin;* and *Forgetting Sarah Marshall*), he has never played the role of an action hero. Lewis, meanwhile, is black, was raised in poverty by a single mother in Lakeland, Florida, and was a major NFL star at the time, and hence a visible representative of hegemonic masculinity. The joke turns on the premise that despite the seemingly unbridgeable gaps separating affluence from poverty, white from black, icons of masculinity from the average guy, Rudd and Lewis are improbably buddies. Their friendship goes back to the cradle, as Rudd explains in the first ad in the series: "Oh, man, Ray and I have known each other our whole lives. We grew up together. Best friends!" The rest of the campaign shows the two friends playing the video game, engaging in verbal dueling, boasting, and performing other acts that characterize a certain kind of friendly masculine competition.

This ad campaign was widely praised. It was also characteristic of a fantasy *Madden* works hard to create and maintain – the association of "regular guys" like Rudd with the NFL, where "the most extreme possibilities of the male body" are celebrated.[1] Though fanciful and light-hearted, the campaign presents the pleasures to be enjoyed by imaginatively taking that leap. In this chapter, I want to explore the strategies by which *Madden* promotes itself. In my analysis of these efforts, I identify a preferred mode of engagement promoted by *Madden* and its corporate partners, which I call the *Madden* imaginary. I trace the key elements of this fantasy, emphasizing the connections to contemporary football's formation of hegemonic masculinity, as well as the strategies by which the game addresses deep cultural anxieties about the security of that formation.

Madden demands critical scrutiny in part because it has been an unparalleled commercial success: since its launch in 1989, the games have sold more than eighty-five million copies and generated more than three billion dollars in revenue.[2] *Madden's* cultural imprint, however, extends far beyond these sales. The game is regularly featured on ESPN's preview, highlight, and analysis programs and holds a key place in the NFL's own promotions, so the game is well known even among those who have never played it. *Madden NFL* is one of the best-selling video games of all time, but perhaps more significantly, it is the unrivaled video game simulation for the most popular sport in the United States. The NFL dominates the sport-media complex in the United States, drawing the eight largest television audiences of any kind in 2012, including the largest U.S. audience in the history of television to its 2012 Super Bowl. But the league's dominance extends beyond television. Football has been a site of innovation for media marketers attempting to connect with fans in new ways. In the contemporary "crowded marketplace" of sport media, the NFL and its media partners have forged flexible marketing strategies to reach consumers as their media consumption habits change with technological, cultural, and economic shifts. As sports marketing experts Irving Rein, Philip Kotler, and Ben Shields explain, sports marketers are confronting a "new era," where "all fans are elusive; all fans are in play. Competitors are engaging in an all-out battle for the money, time, and attention of fans. Sports decision makers are facing a new level of competition, a race to survive in a crowded marketplace, and a struggle to define, attract, and retain the ever-elusive fan."[3] EA Sports, which develops and markets the *Madden NFL* franchise, has been especially innovative in developing ways first to connect with consumers and then to deepen that connection. These arrangements have been mutually beneficial. In addition to collecting a substantial licensing fee, the NFL finds new ways to get its brand before audiences, while extending the length of their engagement with the brand. Meanwhile, EA Sports profits from its role as the primary provider of a realistic NFL simulation.

During the same period, as these technological changes issued new challenges and opportunities to reach consumers, anxieties about the shifting and contested meaning of white masculinity have subtly changed the cultural landscape. Various political and cultural developments have challenged the "identity politics of the dominant" and produced a spate of cultural responses across media, including literature, cinema, television, and new media.[4] These anxieties have also emerged in texts about elite sport. Norman Denzin asserts that it is "only a slight exaggeration to conclude that sports in all facets is the most significant feature of contemporary racial order."[5] Thus, we should not be surprised to find concerns about a shifting racial order concentrated around sport. In American football, where white men last constituted a majority of players in the early 1980s, the anxiety has been particularly acute. Black players have been a majority since 1983, and approximately two of every three NFL players are black currently.

Of course, the disproportionate representation by blacks on the field does not extend to the coaching, management, and ownership. All of the league's majority owners are white, and black coaches and general managers are still rare (though not as scarce as they used to be), and many of the league's most visible stars are white. Nevertheless, perceptions of black dominance in football's most public and glorified roles have been widespread. The anxieties this shift produced were evident early on, when in January 1988 CBS football analyst Jimmy "the Greek" Snyder famously shared his opinion that in the NFL, "all the players are Black." In an impromptu interview with a television reporter, Snyder worried that, should blacks enter coaching, "like everybody wants them to, there's not going to be anything left for the White people."[6] The football analyst also suggested that blacks enjoyed a genetic advantage over whites due to slave breeding practices. Snyder's comments were roundly criticized as insensitive at best (he had made the comments on Martin Luther King Jr.'s birthday) and racist at worst.

Snyder was immediately fired by CBS, but attempts to explain the supposed disappearance of white athletes with racialist arguments did not leave the mainstream with Jimmy the Greek. In 1997 a *Sports Illustrated* cover asked "Whatever Happened to the White Athlete?" and suggested that most young whites had accepted a sense of race-based inferiority and were slowly turning away from participation in mainstream sports like football. Florida State football coach Bobby Bowden lent support to the (apparently) widely shared belief that "an athlete is an athlete, but, dang it, there just seem to be more black athletes than white." Critical scholar Kyle Kusz has examined how the supposed "disappearance" of white athletes from mainstream sports has helped prompt the emergence of extreme sports such as those portrayed in ESPN's "X-Games" and films such as *Dogtown and Z-Boys*. Such entertainments, Kusz argues, express the concerns of a "white male backlash politics" and craft narrative spaces where hope of a "remasculinization" of white identity is explored through sport. White males have been re-centered in narratives about sport in the midst of what David J. Leonard calls "a metamorphasizing sports world – rising player salaries, increasing visibility of black athletes, especially as stars, greater corporate and media interest in sports."[7] But in important hypermasculine athletic rituals such as those staged by the NFL, black bodies remain central, and if anything are growing in visibility.

Black players are increasingly visible in *Madden* as well. *Madden*'s rosters reflect the roughly two-to-one ratio of nonwhites to whites on the NFL rosters, but the stars celebrated by the game have been even less likely to be white. Since the release of *Madden NFL 2001*, when the game's namesake, former coach and commentator John Madden, was replaced on the cover, fourteen NFL players have been featured as cover athletes. Only three have been white. Since 2000 sixty-four players, excluding kickers and punters, have been assigned an overall skill rating of 99 or 100 by *Madden* designers. Only thirteen of those players have been white.

Anxieties about white "disappearance" from hypermasculine spaces have not led to a devaluing of hypermasculine aggression and the link to male power it carries. As bell hooks notes, "Showing aggression is the simplest way to assert patriarchal manhood. Men of all classes know this." As black men's visibility in U.S. media culture has grown, their association with patriarchal aggression has become increasingly normalized, especially through hypermasculine sporting productions such as the NFL. Here, as well as in other cultural locations, black men have been routinely presented as "poster boys of brute patriarchal manhood." As Patricia Hill Collins summarizes, "Recognizing that black culture was a marketable commodity, [commercial media interests] put it up for sale, selling an essentialized black culture that white youth could emulate yet never own."[8] This multicultural patriarchal marketing is the focus of this chapter: How has *Madden* navigated the difficult terrain of contemporary racialized masculinity that lies at the heart of its appeal to consumers? By what strategies does *Madden* present itself as a means to realize fantasies of masculine power and aggression when that fantasy is often coded black, and while deep anxieties about black masculinity continue to circulate in and beyond sport?

This chapter will trace a set of complex hypermasculine fantasies offered to *Madden* gamers through EA Sports' promotional strategies. In these fantasies, which collectively constitute the "*Madden* imaginary," the aggression, strength, and skill of NFL players, as well as the control and authority of NFL coaches and organizations, can accrue to gamers. This combination reflects deep ambivalences in contemporary football fandom, where the affirming possibilities of masculine strength, speed, and physical skill exist uncomfortably alongside the widely expressed suspicion that whites are "disappearing" from football and other hypermasculine sports and frequently expressed disgust about the perceived undisciplined selfishness of the contemporary black-dominated game. The *Madden* imaginary offers fans a way to bridge this difficult tension in football fandom by offering parallel fantasies of embodiment and managerial control. In the pages that follow, I trace three key components of the *Madden* imaginary: EA Sports' relentless and widely publicized pursuit of an utterly convincing simulation of the contemporary NFL, the measures by which EA Sports articulates and strengthens links between the *Madden* experience and the hypermasculine culture of the NFL, and the promotion of fantasies in which gamers control and direct those bodies as managers, and even as creators.

THE REAL THING

In many video games, elements of realism coexist with aspects of fantasy, even otherworldliness. As Ian Bogost argues, other video games engage audiences via a set of procedures that works to create a procedural rhetoric that may be shaped to confront or challenge dominant ways of thinking and acting in the world. In most sports simulations, however, realism is a relentlessly pursued goal. *Madden NFL* is one of the most highly developed examples of this general tendency. From its earliest conception, *Madden* was developed with the goal of achieving as "realistic" a depiction as possible. The game is the brainchild of Trip Hawkins, a passionate football fan and Strat-o-Matic football devotee. As a teenager, prior to his interest in digital games, Hawkins had attempted to launch a mass-market board game based on Strat-O-Matic. The game that the adolescent Hawkins tried to bring into the mainstream was a statistics-based contest similar to fantasy football. It was played by a small subculture of hard-core sports fans, but Hawkins's attempt to bring the game into the mainstream failed because not enough fans were willing to perform the required math to play the game. As ESPN's Patrick Hruby summarizes, "'Strat-O-Matic' was too hard. Players had to crunch too many numbers, obliterating the necessary suspension of disbelief."[9]

Fortunately, by the time Hawkins graduated from Harvard, the possibility of computerized gaming made that problem solvable, allowing for a football simulation with a lot of number crunching that could be performed by computers without the gamer's effort. Hawkins became an early innovator at Apple computers before leaving in 1982 to form Electronic Arts (now known as EA) and again set out to create a convincing football simulation, this one in a digital format. An early innovation pioneered by Hawkins's EA was the inclusion of digital renderings of active star players. In "One-on-One: Dr. J v. Larry Bird," EA pioneered the effort it would ultimately apply as a formula for its most successful franchises – bringing recognizable sporting stars under the virtual control of gamers and making it possible for fans to embody not merely generic athletes, but stars. Henceforth, EA would work not so much to invent digital athletes as to capture with their renderings a likeness, personality, and skill set already known to the gamer.

Though the launch of *Madden* began without a license to use the actual names of teams or players, it strove to replicate football in other ways. In consultation with former NFL coach and then television analyst John Madden, the developers included plays developed by actual NFL teams that gamers could implement at home. The process was exhaustive. One early developer, Joe Ybarra, remembers of his early research into NFL plays: "We spent hours just learning blocking schemes. . . . By the third year of the project, I could watch pro football on TV and tell you what was going to happen when the players were still lining up."[10] In 1993 EA, now having launched a separate division (EA Sports) to handle sports games, was able to negotiate a license from the NFL, allowing developers to construct digital versions of entire NFL rosters. In the quest for realism, there is perhaps no more important feature than access to the names and likenesses of those who play the game. The license has also allowed the *Madden* franchise to outmaneuver rivals in the video game market.

For example, in 2005, under pressure from challenges from ESPN's line of video games, EA Sports negotiated an exclusive arrangement with the NFL, which drove the competition out of business. ESPN, perhaps concluding that it was best to join what one cannot beat, then entered an agreement with EA Sports to integrate brands. ESPN now features *Madden* simulations extensively in its analysis programming and its showcase program, *SportsCenter,* in which digital renderings of ESPN anchors, reporters, and commentators offer commentary during gameplay. Such blurring not only expands the visibility of *Madden,* but also establishes the game as a reliable simulation imbued with a pseudoscientific reliability as a tool of analysis and prediction.

Part of what makes a *Madden* simulation plausible is the game designers' remarkable commitment to gauging multiple components for each player represented in the game. Numerical rankings determine each digital football player's skill level, and the process of establishing player skill levels is exceptionally data intensive. As the game's ratings designer, Donny Moore, explains, this attempt "to portray each player and their on-field performance as accurately as possible" requires consulting a number of publications that cater to hard-core fan interest in football scouting, including "War Room" reports published by the *Sporting News,* an online statistical research site for fans called "Football Outsiders," as well as professional scouting services. Moore stresses the importance of these sources in providing an "accurate" rendering of player performance. He says, "I try to use all that information and accumulate as much information as possible."[11]

Madden rates players by speed, acceleration, and strength and by position-specific and less tangible skills such as awareness, catching, and throwing accuracy. Every player is rated on a 100-point scale, and ratings are adjusted regularly throughout the season to reflect changing evidence of these abilities, updates that are then made available for the growing legion of online players. As Moore explains to Tom Bissell in a feature for ESPN's affiliated *Grantland* site, "'I'm doing a roster update right now with the Packers. I want to make sure that the latest information is in there. Of course, Charles Woodson is a 95 overall corner, and he's a top corner, probably one of the top five in the game, but he just had a terrible game against the Giants. I want to make sure that that's reflected in the update.' With a keystroke Moore knocked down Woodson's offensive awareness by one."[12]

Many fans of the *Madden* franchise understand the game's claim to statistical realism to be a key feature of the game's success. As Andrew Baerg has documented, some fan debates about the realism of *Madden's* statistical simulations for NFL players go so far as to include online discussions about strategies for tweaking the game's calculations to achieve a more satisfying level of realism. Baerg notes a gendered element in this supposedly objective quest to better authenticate the *Madden* experience – what he calls "quantitative realism." He concludes from the fans' posts that their commitment to pursue a program of rational quantitative edits (a process described by the fans as "scientific") represents an affirmation of their manliness: "To act scientifically is to perform their masculinity in dominating the technological environment in which they live." Baerg acknowledges that this manly commitment to scientific mastery has a long history in football, dating back to the game's popularizer, Walter Camp. Thus, he concludes, "It is precisely the scientized expression of this kind of power over the number that serves as an epistemological frame through which *Madden* users interpreted the game's lack of realism. The technological manifestation of the sport of football in the medium of the digital game serves as a prime exemplar of the nexus of gender, science, and mediated sport."[13]

EA Sports makes no secret of game developers' relentless pursuit of realism. Indeed, revealing the effort is a key part of the marketing. The company distributes a promotional documentary titled *The Making of "Madden"* that details the painstaking and highly technical methods used to achieve a believable simulation. The documentary showcases the motion-capture technology that the game uses to construct digital players. In the film NFL players perform for game designers. There is no mention that the vast majority of motion capture is done using a single actor who mimics NFL players.[14] The experience of the "authentic" NFL is the key feature of the promotion's discussion of the game's use of motion-capture technology. As Anthony Stevenson, senior product manager for *Madden NFL 11*, explains for the *Making of "Madden 11"* promotional video, "It's really crucial that our animations reflect what a player really looks like on the field doing a certain move, whether it's cutting or juking or a wide receiver tapping his toes to stay inbounds.... [W]e've got a very, very intensive motion capture process where we actually take NFL athletes, we bring them into a studio, we have sensors all over their body and we just capture every single movement and let that sort of manifest itself inside the game."[15] Stevenson's choice of words here also helps to convey the sense of realism. Motion capture, in his account, simply manifests inside a computer, as if no engineering were needed beyond the simple act of capturing the motion and then relocating it to the game.

In another example of this commitment to the feel of authenticity, *Madden NFL 13* boasts a feature it calls "the Infinity Engine," which produces "physics you can feel." As EA Sports' promotional material explains, the engine "augments animation based on contact with other players or objects on the field," thereby adding "more variety, emergence, and an extra level of authenticity to *Madden NFL 13*."[16] A team of engineers (including three holding doctorates in physics) worked to "ensure that no two plays ever look or feel the same."[17] A designer admits that "I wish I could tell how your interactions will play out but due to the emergent nature of the system," such inauthentic predictability is the very thing targeted for elimination by the engine, so "I just can't."[18] The result of this labor, readers are assured, is a more true-to-life experience, in which each play is potentially unique.

EA Sports' relentless pursuit of realism is a key element of the game's marketing. In constructing the *Madden* imaginary, designers and marketers stress the importance of a "deep" virtual experience in which various aspects of the imaginary can coexist within a particularly constructed virtual world. Capturing the feel of NFL competition (minus the physical dangers) is vital to unlocking the potential pleasures *Madden* offers to the average guy who represents the franchise's target market.

MAKING GAMING MANLY

The union of hypermasculine sports such as football and the audience of gamers is not as comfortable or natural as *Madden*'s financial success might suggest. Indeed, EA Sports has worked carefully to foster a close and easy connection between *Madden* and recognizable codes of hegemonic masculinity, a process pursued primarily by connecting the gamers with the hypermasculine culture of the NFL. In this section, I shall illustrate some of the ways the game celebrates male aggression through *Madden*'s game features, the television series *Madden Nation*, and the *Madden Challenge*. I also examine efforts to build connections between gamers and active NFL athletes through the celebrity tournament *Madden Bowl*.

As Rich Hilleman, a developer who worked on the earliest incarnations of *Madden*, recalls, "Before *Madden*, jocks did not play video games." In fact, when *Madden* was released in 1989, "somebody playing games was more likely to get made fun of on ESPN than get featured on there."[19] In the years that followed, *Madden* joined a broader trend of hypermasculine, violent video games that included first-person shooters such as 1992's *Wolfenstein 3D* and 1993's *Doom*. In the 1992 version of *Madden*, players suffer frequent injuries, and severe injuries are frequently depicted. In such instances, an ambulance is summoned to transport the player off the field (and presumably to medical care), often driving over injured players for humorous effect.

Though such early arcade-style features were phased out as the pursuit of realism intensified, hypermasculinity was later promoted by other means. For example, enhancing control of the athletic movements of *Madden*'s digital athletes has been one of the game's constant features. Most generations of *Madden* offer improvements or new features for on-field control. The "Playmaker Tool," the "Hit Stick," the "Truck Stick," and the "Weapons" features allow gamers greater control over players' ability to inflict damage on an opponent or exercise skill moves with greater proficiency. Promotional material for *Madden 11* boasts that its "dual stick control" makes it possible for gamers to "run through holes, break tackles, and explode in the open field."[20] An ad campaign for *Madden NFL 08* with the tagline "How Does It Feel?" exploits this fantasy as well. In this campaign *Madden* offers up digital renderings of athletic skill and power and invites gamers to imagine themselves inhabiting the agile, fast, and powerful bodies that populate NFL rosters.

A different strategy to link the video game to aggressive styles of masculinity is employed in the ESPN reality series *Madden Nation*. The program, a product of the cooperative agreement between EA Sports and ESPN, conducts an eight-player *Madden* tournament involving contestants who also travel the country on a tour bus. The finals are conducted in Times Square on the ESPN Zone restaurant's public screen. Participation in *Madden Nation* is governed by an aggressively hypermasculine code of conduct. Contestants are known to the audience and to the other participants by nickname, most of which reference physical destructiveness ("Mad Dog," "UFC Champ"), celebrity ("Hollywood," "KStarr"), or macho boasting and "trash talk" ("The Gift," "Dynasty," "Yomama"). During competition, it is common for opponents to be shown engaging in the kind of aggressive trash talk one might expect from athletes themselves. The program's performers are overwhelmingly male (the one female contestant in the entire four-year series, Sheila Barger, a.k.a. PG-13, was eliminated in the season's first episode).

This aggressively masculine code of conduct governs the *Madden Challenge* as well. The *Challenge,* an online tournament involving thousands of participants and sponsored by Virgin Gaming, cumulates in a live event in Las Vegas, where finalists compete for "life changing prizes."[21] The *Challenge* includes a prominently featured space on its website, where contestants can upload "Trash Talk Videos" and engage in masculine boasting. Qualifiers for the semifinals are commemorated on the *Challenge* website in profiles that resemble NFL player trading cards.

Connections to the hypermasculine possibilities of the *Madden* imaginary are produced in other ways as well. Each gamer on *Madden Nation* is affiliated with a star NFL player, and the meeting between the two is a regular feature. At these meetings, the NFL player typically expresses his enthusiasm for the video game, offers advice and best wishes to the gamer, plays a game of *Madden* with the gamer, and presents the gamer with a replica jersey with the player's name and number. The gamer then wears the jersey for the rest of the season, "representing" the player with whom he is partnered. This explicit connection between NFL players and the gamers is a well-established feature of EA Sports' promotional strategy.

Since 1995 EA Sports has worked in partnership with two other media content providers to stage the *Madden Bowl,* a celebrity tournament in which active NFL players and other celebrities compete via the *Madden* platform. Proceeds from the inaugural event went to an NFL charity; the winner, running back Reggie Brooks, pocketed a thousand-dollar prize; and the video game had a successful promotion. At the 2011 Super Bowl, a new format was implemented for the now-annual event, in which teams of three celebrity gamers competed for the prize. The 2012 event was carried live on ESPN3.com, with ESPN personalities Trey Wingo and Michelle Beadle co-hosting the ninety-plus-minute program. The two co-hosts engaged in lighthearted banter and played out a running gag in which Wingo denigrated and ostracized his female co-host, going so far as to hold private conversations with athletes on the set while

Beadle sat nearby in faux-humiliated silence. Wingo and Beadle also openly acknowledged her "subservient role" throughout the program.

That display of masculine dominance underscores a macho fantasy constructed around *Madden,* in which viewers see skilled *Madden* gamers who are not emasculated nerds or geeks, but are instead NFL players – hegemonic icons of athletic masculinity. The ESPN commentators hosting the event work to blur the lines between the athletes' exploits during *Madden* play and their work on the field in actual NFL games. For example, when a team of gamers including NFL quarterback Tim Tebow won the 2012 *Madden Bowl* with a game-winning touchdown in the final minute, Trey Wingo referenced Tebow's recent on-field heroics: "You're telling me a team with Tim Tebow on it did something weird in the fourth quarter to win the game? That never happens!"

The *Madden Bowl* does more than simply promote the video game franchise. It also works, like *Madden Nation,* the *Madden Challenge,* and each year's new gameplay features, to mark the game as a male endeavor. By making connections between the hypermasculine athletic prowess of actual NFL players with the *Madden* video game, athletic achievement and the exploits of gamers are connected, and the lines separating them are blurred. Thus, the imagined connections between gamers and the players they digitally inhabit through gameplay are strengthened and legitimated. The ESPN broadcast of the *Madden Nation* and the *Madden Bowl* also works to mark *Madden* as a space where women are, if not unwelcome, then relegated to a subservient role. Interestingly, this move works to reinforce the supposed realism of the *Madden* imaginary, because in football women occupy similar roles – present but secondary, relegated to being cheerleaders or sideline reporters.

CONTAINING BLACK MASCULINITY

As the section above details, *Madden* promotes a close affiliation between gamers and the hypermasculine culture of the contemporary NFL. In these moves, the *Madden* imaginary encourages gamers to imagine themselves embodying NFL players and promotes an aggressively macho (and almost exclusively male) culture of *Madden* gamers via the

Madden Bowl, Madden Nation, and the *Madden Challenge.* But the game not only seeks to exploit connections to the contemporary game, but also constructs elaborate scenarios in which gamers can enjoy the fantasy of controlling these athletes in various ways. In this section, I explore some of the features offered by *Madden* that invite fans to deploy a managerial outlook, exercising imagined control over NFL athletes.

In 1998 *Madden NFL* introduced a new option called "Franchise Mode," in which gamers were invited to guide a team through multiple seasons, drafting new players and orchestrating trades. Gamers experiencing *Madden* through Franchise Mode act as head coach: making substitutions, organizing practices, and calling plays in addition to personnel matters usually the responsibility of general managers. As the Franchise Mode developed, new features were added, and the name was changed for a time to "Owner Mode." Through the eyes of the owner, gamers still exercised the duties of the coach and general manager, but also took on other concerns, such as the price of tickets and concessions and even the design of the stadium. *Madden NFL 12*'s Franchise Mode included new features to deepen the experience of scouting new talent and negotiating free-agent deals. The next year *Madden NFL 13* rolled the Franchise Mode into the "Player Career" heading, which offers the options previously available as Franchise Mode under the heading "Coach Career." The shift was part of a move by game designers to build a more integrated universe, where other options of the game could be coordinated with the features from Franchise Mode.

Other opportunities for engaging in this fantasy of control are made available as a way of promoting the game. *Madden Ultimate Team, Madden Superstars,* and *Madden Social,* offered online and through the social networking site Facebook, invite users to compete in a variety of ways by manipulating virtual "rosters" of NFL players. In *Ultimate Team,* participants compete by assembling through trades the "ultimate team" of player cards with scores reflecting their value. *Superstars* is a Facebook site that organizes what is essentially a fantasy league game, in which contestants seek to build the most valuable and productive virtual team through drafts and trades and by playing some players while "benching" others. *Madden Social* combines elements of the *Ultimate Team* game with a scaled-down version of the console-based game. Clearly, these attempts to reach online users are driven by a desire to imagine new ways to connect with audiences. However, the *way* these games seek to establish these connections is significant. Each offers a similar fantasy of controlling a roster of NFL players and deploying it strategically.

This desire to extract the maximum productivity from NFL players is perhaps most developed in the long-standing feature of the *Madden* console game "Superstar Mode." This mode allows gamers to experience a career from a player's perspective, but even here extraordinary fantasies of control are made available. The popular "Create a Player" option allows gamers to literally build a digital avatar to their specifications. Gamers select the type of player they would like to create. For example, one can select a quarterback from several options, choosing between a "Pocket Passer QB" or a "West Coast QB," among several other choices. Gamers can also design the size and shape of the player's head, his hair color and type, arm size and definition, as well as skin color. Gamers can choose the player's physical abilities, including "strength" and "speed," and can manipulate ratings for more elusive features such as "toughness," "awareness," and "confidence." Gamers can even outfit their players in their favorite brand of athletic shoe and select the team for which they will play.

This fantasy of control over athletes, the ability to deploy them strategically, invest in promising players, and trade or cut unproductive players, constructs an important feature of the *Madden* imaginary. As owners/coaches/general managers (distinctions that the game usually blurs), gamers imagine themselves controlling and directing NFL players. Of course, this fantasy of control is not limited to black athletes. White and black NFL players are subject to these fantasies of control. However, as I have argued elsewhere, such fantasies (which are also available in fantasy leagues and the mediated coverage of the NFL draft) have enjoyed remarkable growth and mainstream popularity during a period of widely expressed anxiety about a black-dominated game.[22] They represent a significant shift in the representational strategies for hypermasculine sports such as football. As whites are understood to be "disappearing" from elite football, the terms of engagement have subtly shifted. Where players were once hailed as icons of masculinity, emerging representational strategies additionally position them as commodities.

CONCLUSION

The NFL has grown in popularity during the past several decades, a period when white masculine domination has faced a widespread set of challenges. As a number of scholars have noted, professional football has served as a bulwark against the insurgencies from feminists and other groups. Michael Messner has argued that "sports as a mediated spectacle provides an important context in which traditional conceptions of masculine superiority – conceptions recently contested by women – are shored up." Less attention has been paid to the factor of race, which complicates this process significantly. Though NFL players arguably represent widely accepted models of hegemonic masculinity, the black-dominated game nevertheless stirs deep anxieties. Consider, for example, the marketing survey Q Scores, which rates the relative popularity of recognizable athletes in the United States.[23] In repeated surveys, black NFL players dominate the list of the most unpopular athletes, while white NFL stars are consistently among the most popular.

There are, of course, exceptions to this general tendency, but the trend, in combination with recurrent tropes extolling the virtues of the old-school game (when blacks were not so visible) while bemoaning the current game as the triumph of style over substance, suggests a deep anxiety among many members of the mainstream with the demographic shifts in pro football. The performance of hegemonic versions of masculinity remains central to the formation of cultural and political power in the United States, and public enactors of these norms are as vital as ever. The fact that, in the NFL as well as in other spheres, those playing those roles are often black creates a profound challenge.

Madden NFL offers fans a way to bridge this difficult tension by offering parallel fantasies of embodiment and managerial control. *Madden* gamers are invited to imagine themselves inhabiting the hypermasculine bodies of NFL athletes – virtually executing plays as an NFL player. As I have detailed above, the various promotional strategies undertaken by *Madden* seek to establish a clear connection between the aggressive, hegemonic culture of the NFL and the *Madden* imaginary. These engagements with NFL players, however, also invite gamers to imagine themselves controlling and directing players, disciplining them, training them, even buying and selling them.

Of course, there is no guarantee that audiences will necessarily engage the video game on the terms endorsed by the *Madden* imaginary. Though commercial media producers promote particular kinds of pleasure, this does not determine the uses audiences will make of their products, nor does it make impossible alternative, even subversive, forms of pleasure that might be derived from popular media. Nevertheless, it is important to carefully document and theorize the kinds of pleasures endorsed by these producers. Doing so potentially opens spaces for alternative readings and helps us to better understand what meanings powerful commercial and cultural forces are attempting to promote.

Madden's notable commercial and cultural significance is due not only to the franchise's remarkable integration with some of the most recognizable and profitable brands in the contemporary sporting economy, but also to the sophisticated fantasies it constructs, which navigate the complex, shifting terrain of racialized masculine dominance. *Madden* offers gamers the virtual embodiment of NFL athletes, experiencing the thrill of dominating opponents in the arena of hypermasculine conquest. At the same time, it affords gamers the fantasy of controlling virtual NFL rosters – developing athletic talent as a commodity and deploying it in virtual competition.

NOTES

1. Michael Messner, *Out of Play: Critical Essays on Gender and Sport* (Albany: State University of New York Press, 2007), 72.

2. Patrick Hruby, "The Franchise: The Inside Story of How *Madden NFL* Became a Video Game Dynasty," http://sports.espn.go.com/espn/eticket/story?page=100805/madden.

3. Irving Rein, Philip Kotler, and Ben Shields, *The Elusive Fan: Reinventing Sports* in a Crowded Marketplace (New York: McGraw-Hill, 2006), 8.

4. See, for example, Robyn Wiegman, *American Anatomies: Theorizing Race and Gender* (Durham, NC: Duke University Press, 1995); Sally Robinson, *Marked Men: White Masculinity in Crisis* (New York: Columbia University Press, 2000); Brenton Malin, *American Masculinity under Clinton: Popular Media and the Nineties "Crisis*

of Masculinity" (New York: Peter Lang, 2006).

5. Norman Denzin, "More Rare Air: Michael Jordan on Michael Jordan," *Sociology of Sport Journal* 13, no. 4 (1996): 319.

6. "Jimmy 'the Greek' Snyder Slurs Black Athletes in King Interview; Apologizes," *Jet,* February 1, 1988, 18.

7. S. L. Price, "Whatever Happened to the White Athlete?," *Sports Illustrated,* December 8, 1997, 34; Kyle Kusz, "'I Wanna Be the Minority': The Politics of Youthful White Masculinities in Sport and Popular Culture in 1990s America," *Journal of Sport and Social Issues* 25, no. 4 (2001): 390–416; Kyle Kusz, "*Dogtown and the Z-Boys:* White Particularity and the New, *New* Cultural Racism," in *Visual Economies of/ in Motion,* edited by C. Richard King and David J. Leonard (New York: Peter Lang Press), 135–163; David J. Leonard, "To the White Extreme: Conquering Athletic Space, White Manhood, and Racing Virtual Reality," in *Digital Gameplay: Essays on the Nexus of Game and Gamer,* edited by Nate Garrelts (Jefferson, NC: McFarland, 2005), 116.

8. bell hooks, *We Real Cool: Black Men and Masculinity* (New York: Routledge, 2004), 51; Patricia Hill Collins, "New Commodities, New Consumers: Selling Blackness in a Global Marketplace," *Ethnicities* 6, no. 30 (2006): 298.

9. Ian Bogost, *Persuasive Games: The Expressive Power of Videogames* (Cambridge: MIT Press, 2007); Hruby, "The Franchise."

10. Hruby, "The Franchise."

11. *"Madden NFL* 12 Interview: Donny Moore – Scouting Talent," http://forum .ea.com/eaforum/posts/list/5240087 .page.

12. Tom Bissell, "Kickoff: *Madden NFL* and the Future of Video Game Sports," http://www.grantland.com/story/_/id /7473139/tom-bissell-making-madden-nfl.

13. Andrew Baerg, "It's (Not) in the Game: The Quest for Quantitative Realism and the *Madden* Football Fan," in *Sports Mania: Essays on Fandom and the Media in the 21st Century,* edited by Lawrence W. Hugenberg, Paul M. Haridakis, and A. C. Earnheardt (Jefferson, NC: McFarland, 2008), 228.

14. "Eye to Eye: Kenney Bell," *60 Minutes,* January 21, 2008, http://www .cbsnews.com/video/watch/?id= 3734987n.

15. Christopher Erb, Michael Herst, and Kirk Langer, producers, *Making of "Madden NFL"* (Spike TV Video, 2010), DVD.

16. Victor Lugo, "Infinity Engine Developer Blog," June 3, 2012, http://www .easports.com/madden-nfl/news/article /infinity-engine-developer-blog.

17. "Feature: Infinity Engine," http:// www.easports.com/madden-nfl/feature /gameplay-part2.

18. Lugo, "Infinity Engine Developer Blog."

19. Hruby, "The Franchise."

20. *"Madden NFL 11,"* http://www.ea .com/au/madden-nfl.

21. "How It Works," http://virgin gaming.com/tournaments/challenge -series/how-it-works.html.

22. Thomas P. Oates, "New Media and the Repackaging of NFL Fandom," *Sociology of Sport Journal* 26, no. 1 (2009): 31–49.

23. Messner, *Out of Play,* 54. See http:// www.qscores.com/Web/Index.aspx.

THREE

Neoliberal Masculinity: The Government of Play and Masculinity in E-Sports

Gerald Voorhees

> We're at a point where only about forty people in the U.S.
> can make a living playing video games. I'd like to get it to a
> hundred. I think we're a year or two away from that.
>
> SUNDANCE DIGIOVANNI, quoted in Richard Nieva,
> "Video Gaming on the Pro Tour for Glory but Little
> Gold," *New York Times*, November 28, 2012

While scholars have begun to investigate the professionalization of gaming, I take it on only to the extent that it is an exemplary site for thinking about the sportification of digital games, a broader sociocultural phenomenon that emerges at the juncture of neoliberal rationality and distinct – often competing – constructions of masculinity circulating in contemporary Western culture. Indeed, the sportification of digital games has led to the creation of national leagues, international tournaments, and corporate-sponsored teams of professional cyberathletes, but it is not rooted in these institutions or in the professionalization of players; rather, they are both effects of the hegemony of the sportive mentality. The games are objective things defined by protocological affordances and constrains, but their status as sport and the practices constituting the process of sportification are a result of the meaning attributed to them by player and fan communities.[1] In this chapter I examine the cultural implications of the figuration of digital games as sports, often called e-sports, focusing on the production of an intelligible subject position at the nexus of neoliberalism and masculinity.

I argue that the sportification of digital games, and the subsequent professionalization of gaming, contributes to the production of a form of subjectivity characterized by technological mastery, economic rationality, and the ennobling of violence, which I term *neoliberal masculinity*. Approached with a sporting mentality, digital gaming exemplifies the constant and ubiquitous cost-benefit analysis demanded by neoliberal rationality. Play for the sake of play is displaced by the calculative economization of action. E-sports also stage the spectacle of serious and consequential competition through games, bringing together often antagonistic notions of masculinity centered on technological mastery and intelligence, on the one hand, and physical mastery and violence, on the other hand, with childish play. This chapter will show that as e-sports, digital games facilitate the meeting of neoliberal and masculine ideologies, and the sportification of games also produces a historically specific form of masculinity that cannot be disentangled from neoliberalism. This neoliberal masculinity accepts both the violent and the reasoning aspects of masculinity, not because professional gaming is a space of tolerance for difference but because the context of the e-sport makes their union an appropriate strategy for success.

The chapter proceeds from a discussion of the scholarship that informs this argument before turning to analyze the cultural practices of sportive gameplay. I begin with an overview of Michel Foucault's theorization of neoliberal governmentality, which elaborates the means by which persons turn themselves into subjects. In this formulation, cultural knowledges determine human action to the extent that they constitute a grid of intelligibility, a field of possible actions upon which subjects plot their own course. I then examine, with the aim of bridging the concepts, how play and masculinity are understood in relation to sport. I first trace the relationship between several distinct notions of play, distinguished by the extent to which they are rationalized, in games and sports. I then look to research on cultural constructions and transformations of masculinity, focusing on the muscular, sporting body that emerged in the early twentieth century. Contemporary research by T. L. Taylor and Nick Taylor on the professionalization of digital gaming as well as my own observations of the North American industry inform my application of these notions to contemporary e-sport.

NEOLIBERAL GOVERNMENTALITY

Neoliberalism is a multifaceted and therefore somewhat problematic construct. In the main, it is used to describe an economic system characterized by deregulation of economic activity, the expansion of the private sector into areas once considered the public domain such as education, incarceration, and warfare, and the opening of new markets to global flows of capital and free trade.[2]

However, as I employ the term, neoliberalism refers to processes simultaneously narrower and more diffuse than this strictly economic sense. As developed through the work of Michel Foucault, particularly in the 1978–1979 lectures at the Collège de France published as *The Birth of Biopolitics,* neoliberalism extends the line of economic theory that stretches from Adam Smith's laissez-faire to the Chicago School into the multifaceted arenas of everyday life. This "massive expansion of the field and scope of economics" potentially extends into "everything for which human beings attempt to realize their ends, from marriage, to crime, to expenditures on children, [which] can be understood 'economically' according to a particular calculation of cost for benefit."[3] Everything includes the state, and so "government itself becomes a sort of enterprise whose task is to universalize competition and invent market-shaped systems of action for individuals, groups and institutions."[4] This marks a dramatic shift in (but not by any means a reversal of) the political construct of liberalism, which seeks to safeguard civil society against encroachments from the state. In this neoliberal condition, the boundary between society and government, in the political-juridical sense, becomes porous, as both domains are reconfigured by the influence of economic rationality and, in this way, governmentalized. Foucault introduces the concept of governmentality to explain the shift in his theorizing beyond disciplinary power, which dominates and *subjectifies* people, to also account for the grids of intelligibility that organize and enable *subjectivization,* the process by which persons turn themselves into subjects. While government refers, quite broadly, to the question of "how to be ruled, how strictly, to what end, by what methods, etc.," governmentality refers quite specifically to "the ensemble formed by the institutions, procedures, analyses and reflections, the calculations and tactics that allow the exercise of this very specific albeit complex form of power, which has as its target population, as its principal form of knowledge political economy, and as its essential technical means apparatuses of security."[5] In this formulation, the theorization of governmentality is an analytical move away from the rule of sovereigns (concerned with nation-states) and the domination of disciplines (concerned with individuals), which both fall under the rubric of government but not within the specific domain of governmentality: society.

Neoliberalism is central to Foucault's understanding of governmentality. The ensemble constituting governmentality coheres in the same way that a complex system is more than the sum of its component parts; by virtue of the possible linkages, functions, and transformations – in short the relations of interdependence – between objects within its domain, a system of governmentality maintains the welfare of a population. A crucial but little-examined concept in Foucault's later work, population is grounded in the emphasis of a "common abstract essence" that allows each person to be thought of as equivalent to every other.[6] A population allows for statistical analyses of patterns, trends, and rates. In this way, population is both a "mass" of beings and a "purchase for concerted interventions (through laws, but also through changes of attitude, of ways of acting and living that may be achieved by 'campaigns')."[7] In other words, population is the entry point of economy, not the realm of finance but rather the management of the "imbrication of men and things," into government. It is this conjunction of population and economy that enables the rationalization of the relations between persons, goods, and resources that is essential to the "science of government" and the operation of apparatuses of security.[8] In contrast to the individual focus of disciplinary apparatuses, security is concerned with the employment of tactics that exploit the relations between things in order to create a milieu capable of generating a set of desired effects. By thus "stacking the deck," an apparatus of security endeavors to apply economic principles, by undertaking the "rationalization of chance and probabilities," to manage and cajole the regularities of a population and suppress the aleatory.[9] Governmentality, then, operates at a distance. With a thorough knowledge of the relevant objects and their relations to one another and the population, governmentality is concerned not with direct intervention, but rather with establishing the conditions capable of regulating the economy of a system – economic, political, cultural, and so on.

Studies of governmentality have tended in one of two directions – on the one hand, the examination of purposeful actions on the part of the state and social institutions and, on the other hand, the diffusion of neoliberal rationality throughout everyday life, which turns culture into a form of government (from an analytical standpoint). This necessarily entails an understanding of culture as discursive, or even rhetorical, rather than material. As "a distinctive set of knowledges, expertise, techniques and apparatuses which – through the roles they play as technologies of sign systems connected to technologies of power and working through the technologies of the self – act on, and are aligned to, the social in distinctive ways," culture is coterminous with governmentality.[10] A far cry from Matthew Arnold's notion of culture as the civilized and civilizing qualities cultivated by a social system and even from Raymond Williams's germinal idea of culture as shared symbolic systems, thinking through the relation between culture and governmentality means approaching culture as a "grid of intelligibility."[11] This is to say that people act in accordance with what makes sense in a particular situation given what they know and how they understand the probable effectivity of their possible actions. In this way, power governs; disciplinary technologies and technologies of the self, "which permit individuals to effect by their own means or with the help of others a certain number of operations on their own bodies and souls, thoughts, conduct, and way of being," operate in and according to the same cultural logic.[12]

While Foucault's corpus tackles several technologies of the self, Greek philosophy and Christianity being the most oft cited, this analysis of the professionalization of digital gaming is concerned with neoliberal rationality as a technology of the self utilized to produce *Homo economicus*. "A free and autonomous 'atom' of self-interest who is fully responsible for navigating the social realm using rational choice and cost-benefit calculation *to the express exclusion* of all other values and interests," *Homo economicus* is the figure of the subject suited to the social condition of neoliberalism.[13] Literally, "economic man," and possessing the "common abstract essence" of the population, *Homo economicus* is the sort of individual who navigates the possibilities enabled by the material and cultural milieu in search of the most advantageous probabilities. For this subject, neoliberalism is an ethical imperative to think through the cost and benefit of each potential action with an eye toward developing the self as "human capital," a measure that accounts for both one's inborn qualities and the skills and abilities acquired through life choices.[14]

The professionalization of digital gaming is a means of bringing into being *Homo economicus* by incentivizing, through high-stakes competition, a way of being dictated by rational choice and cost-benefit analysis. As I will argue, it rationalizes the activity of play and legitimates patterns of interaction that are otherwise unacceptable within the dogmas of hegemonic masculinity. By inculcating neoliberalism as a technology of the self in players (professionals as well as the much more numerous hobbyists participating in the cultural zeitgeist), professionalized gaming contributes to the increasing ubiquity of neoliberal governmentality. It is both a sphere of social life once governed by its own norms but now under the sway of neoliberal rationality and a site where subjects are prepared to enact the practices of *Homo economicus* in other fields of life.

PLAY, GAMES, SPORTS, AND MASCULINITY

Despite their several-thousand-year history as situated cultural practices emerging and developing in relation to the societies that play them, sports, like games, have been subject to disproportionately little serious academic scrutiny. However, the scholarly turn to examine popular culture and the practices of everyday life – once largely dismissed as mundane and unremarkable but now recognized as "fundamental to how society itself is made and remade" – has resulted in increased attention to sports.[15] This work is, in some sense, an extension and further specification of the effort to study the relationship between games, play, and culture.

In this section I attempt to find a footing in this literature and establish the relationship between play, games, sports, and masculinity. There is a positive correlation with masculinity all along the continuum of increasingly particular activities that moves from play to game to sports.

I work to bring this into relief, highlighting the increasingly prominent place of physical domination and the accumulation of status in sport. However, I also aim to explicate the distinctions between play and sport in order to foreground the rationalization that accompanies the sportification of games. The experimentation and exploration of gameplay are replaced by the instrumental play of sports in step with the increasingly masculine character of the activity. Though the rationalization and masculinization of sportive play are inextricably intertwined, I attempt to parse them into two succinct discussions, beginning with a discussion of rationality and sport.

From Pure Play to Sportive Play

Play is a multifaceted term, full of contradictory sentiments and impulses; it is in many ways ambiguous.[16] Even setting aside the noun forms of the word, to play in the pure and unconditional sense is distinct from the playing of a game and even further removed from the playing of a sport. I will use the terms *play, gameplay,* and *sportive play* in order to set apart these different activities and in this section work to outline the qualities that distinguish them. In so doing, I hope to establish the relationship between sports and rationality.

Johan Huizinga's seminal anthropological study of play in Western cultures, *Homo Ludens,* was certainly not the first instance where scholarly attention is focused upon play (works by Plato, Kant, and Schiller attest to this), but rather the first effort to develop and elaborate a definition of play for itself.[17] In this regard, Huizinga offers a discussion of play in its "primary significance," as experienced by the player in the moment rather than as an instrument to any particular end. It is this focus that leads Huizinga to his most oft-cited concept, the "magic circle," the idea that play is distinct from "ordinary" life not only because it is limited to a particular time and place but also because it is "disinterested" and "stands outside the immediate satisfaction of wants and appetites." In other words, play is for play and nothing else; if some gain is realized, it is incidental. This puts play squarely in the realm of the mental, or sentimental; it is the intention, attitude, or spirit of play that is essential. Or, as Piaget argues, play is not the action itself but the orientation. It is also worth noting that Huizinga accepts that play adheres to rules, but especially so because it is its rule-bounded nature that makes play constitutive of order and lends it "rhythm and harmony."[18] This is neither a paradox nor a contradiction, as Huizinga does not claim that the point of play is to establish order, but rather that there is more to play – including the culturally constitutive enactment of patterns of activity – than what is intended, a lesson that is fundamental to Roger Caillois's sociological inquiry of play.

Building upon Huizinga's work, Caillois centers rules in order to theorize the movement from pure play to gameplay. He distinguishes *piadia,* characterized by "diversion, turbulence, free improvisation and care-free gaiety," from *ludus,* play structured by "arbitrary, imperative, and purposefully tedious conventions."[19]

The pure play of *piadia* constitutes one extreme of the spectrum of ludic activity and, conceptually, is celebrated by defenders and detractors of modernity alike. It is a vital element in Kant's aesthetics and thus a key to his philosophy of practical judgment. What is present in and thus distinguishes aesthetic judgments from both purely rational and merely animal responses is a "harmonious," "indeterminate" "free-play of Imagination and Understanding."[20] For Kant, this free play is validated by its agreement with reason, but Fredric Schiller takes up Kant's work on aesthetics and argues for a "play impulse" that is not subordinate to but synthesizes the sensual and the rational. Subject to neither the demands of reason nor the impulses of the material body, play has the capacity to transcend both and, "when stimulated by superabundant energy, manifests itself in the free, non-utilitarian exercise of [one's] various facilities."[21] Though his interests are different, Piaget also theorizes a form of pure play. He maintains that only very early in life do people play by blatantly ignoring the precepts of reality, outside of any externally imposed set of expectations or rules that simultaneously serves a socializing function.[22] This type of play is concerned with exploring one's capabilities and experimenting with what is and is not possible; before rules enter the equation, there are only possibility and endless potential.

Where *piadia* is a primal act of "improvisation and joy," the rule-governed gameplay of *ludus* "consists of the need to find or continue at once a response *which is free within the limits set by the rules*" and contributes to the transmission and refinement of culture. To better explain the sociological dimension of gameplay, Caillois invents four classes of games: *agon* characterizes competitions and contests, *alea* defines games of chance, *mimicry* is concerned with illusion and representation, and *ilinx,* finally, is characterized by vertigo and disorientation. Caillois is suspicious of *piadia* play, which he associates with *ilinx* and *mimicry* and further relates to the aestheticized politics of Nazi rallies and argues for a return to a Greek model of civic life, which he characterizes as *ludic* and associates with *agon* and *alea* games. Caillois's advocacy of gameplay is not so much a rejection of Kant's Enlightenment ideals but rather a post–World War II reformulation of them. Intent on preserving the equality and fairness he deemed prerequisite of rational society, Caillois rejects the potential chaos of play and assigns *piadia* a marginal role in social life. Piaget also advocates gameplay over pure play. Where pure play is, in Piaget's terms, an act of assimilating the world with one's own thoughts and imagining, gameplay is a matter of accommodating the social conventions that name specific aspects of reality and define how we are to relate to it. Indeed, Piaget's reflection on play and gameplay is part of a larger effort to understand how children overcome egocentrism in order to become rational, adult subjects. In this formulation, it is not play but gameplay that reconciles the "assimilation of reality to the ego" with "the demands of social reciprocity."[23]

In the social sciences and Enlightenment philosophy alike, we see a concept of pure play tempered with the recognition that sociality requires rules. For Kant and Schiller, play harmonizes with or constitutes this order, but Caillois and Piaget both warn against pure play. They see too much freedom and too little restraint in play. The indeterminate quality celebrated by Kant and the experimental and exploratory potentials lauded by Schiller are regarded, by those writing after World War II's betrayal of the Enlightenment, with suspicion.

Gameplay, rule constrained as it is, provides a degree of certainty and fixedness, a (democratic) center from which we will never stray too far. Jacques Derrida, often cited to lend authority to a notion of play as purely fluidity and indeterminacy, actually theorizes a form of gameplay.[24] He treats play and rules as inextricably intertwined and writes: "The concept of centered structure is in fact the concept of a freeplay based on a fundamental ground, a freeplay which is constituted upon a fundamental immobility and a reassuring certitude, which is itself beyond the reach of the freeplay."[25] This means that play occurs within the rules and therefore cannot rewrite them, which is simultaneously a limitation and a necessary defining quality of gameplay. Foucault works from a similar understanding, focusing not on rewriting the rules but rather on the "moves" available within "truth games," the various fields of human knowledge that constitute the grids of intelligibility that "human beings use to understand themselves" and decide how to act.[26] Although he does not use the term, we get from Foucault a sense of play as the constant negotiation of the different moves made possible by the configuration of the game – for example, as gameplay. The idea that power operates like a game is more than an inconsequential metaphor. Foucault extends the analogy between human action and gameplay when he explains that power can be exercised only on "free" subjects "faced with a field of possibilities in which several kinds of conduct, several ways of reacting and modes of behavior are available." Further, he employs the term *agonism* to describe the relationship between power and resistance "that is at the same time mutual excitement and struggle."[27] This suggests that contestation occurs as multiple, distinct strategies chart different courses through the fields of possible actions. Although gameplay is rule bound, it is far from overdetermined.

Sportive play is a wholly different type of play, not quantitatively more determinate but determined to the last contingency. Allan Guttmann's work on the nature of modern sport, wherein he argues that "spontaneous play is paradigmatically separate from modern sports," is a productive, though also problematic, point of entry to this type of play. To set the stage for his deeply situated, historical analysis of American sports, Guttmann elaborates a typology of playful activities that builds from Huizinga's and Caillois's theorizations of play and games. Moving through a series of dichotomies, Guttmann parses spontaneous play from the organized kind, which he calls games; noncompetitive games from the competitive variety, which he calls contests; and intellectual-oriented contests from those focused on physicality, which he terms sports. Where Caillois and some contemporary scholars insist that "it matters little that some games are athletic and others intellectual," for

Guttmann the marginality of intellectual engagement and centrality of physical, athletic play activity is *the* criterion that distinguishes sports from other contests. This is the first problematic aspect of Guttmann's theorization, and, somewhat strangely, he does not linger on the matter of physicality and instead addresses the contrast between sport, work, and art. Regarding the latter, he argues that unlike artistic activity, sportive play requires no audience and cares nothing for communication. Regarding the former, Guttmann argues that in sportive play, more than any other form of play, there is a more rigid, spatial-temporal separation of play from productive activity.[28] Both of these claims set sportive play in a world apart from everyday life, as if sport were concerned only for itself, and ignore two interrelated and important facets of sport: commercialization and mediated spectatorship. While these are properly characteristics of sports, we would be remiss to ignore their impact on sportive play. Professional athletes are very much caught up in the spectacle of their play and what it communicates to various audiences (fans, coaches, broadcasters, broadcast markets, and others) about their worth. But – and here is a second problematic element in Guttmann's analysis, one in which he problematizes his own claims – amateur athletes of modern sports also play for more than the sake of pure play. They play particular roles in a manner that keeps the rules while aiming for a felicitous outcome, either a win or, even more spectacularly, a record.

From his general definition of sports as "'playful' physical contests, that is, as non-utilitarian contests," Guttmann elaborates seven essential characteristics of modern sport. He identifies secularism, equality of conditions of competition, specialization of activities, rationalization of actions, bureaucratization of rules, quantification of outcomes, and the fetishization of records. Though a more elaborate recounting of these traits should be reserved for another occasion, it is no coincidence that the characteristics Guttmann attributes to modern sport resonate with Max Weber's theorization of the rationalization of capitalist societies. Guttmann's work on sport emerges from the discipline of sociology and emphasizes the contribution of sport to social organization. He does not, however, march in lockstep with Weber. Instead of linking the rationalization of sport to the entanglement of capitalism and Protestantism, Guttmann points to an even more fundamental element of Western modernity: "The emergence of modern sports represents neither the triumph of capitalism nor the rise of Protestantism but rather the slow development of an empirical, experimental, mathematical *Weltanschauung*." In this formula, sport is a "rationalization of the Romantic," which is to say that sport is in fact an expression of the irrational, affective, and even primal qualities repressed by modernity. But this notion of sport as ideology, as false appearance covering the real nature of the experience, privileges the ideal of sportive play over the lived experience of it. In fact, the specialization of activities and rationalization of actions routinizes sportive play. Professionals and amateurs alike practice formations and run plays, maneuvers designed and calculated to achieve a desired effect. In some sports there are nothing but plays; in other sports action occurs between called plays, but that action is never free and spontaneous. Rather, excluding the most novice players, it consists of highly deliberate actions premised on basic principles recognizable to lay spectators and advanced techniques expected by other skilled players. Even if the psychological impact of sportive play is to engage one's base desires, in the actually experienced play, Miller argues, "Sentiment and behavior are codified, supplanting excess and self-laceration with temperate autocritique."[29]

Sportive play and pure play can be located at the extremes of a conceptual continuum of playful activities, with gameplay situated roughly in between. Each of these terms describes not only the increasingly rationalized and rote experience of the activity but the pattern of the activity as well. The "rhythm and harmony" Huizinga associates with spontaneous play, the stratagems of gameplay, and the routines of sportive play each exhibit order and regularity. However, while this order emerges from the unbridled pleasure of pure play and from the agonistic struggles of gameplay, in sportive play it is codified by the norms, traditions, and expectations of the athletic community. This is not to suggest that sportive play is scripted from start to finish. In fact, a key component of athleticism is the ability to respond to the unexpected with skilled improvisation. But it would be a mistake to conflate skilled improvisation with spontaneous experimentation and to overlook the rich bodies of knowledge about sport that issue general guidelines for action even when specific routes and routines are not operating. In short, sportive play is governed by a depth of knowledge of the sport developed through years of study (by players as well as affinity groups, including academic researchers). What distinguishes gameplay and sportive play is the range of viable possible actions, which narrows as games become sports, and the potential effectivity of novel and exploratory acts, which decreases as the field of possible actions enabled by the rules is thoroughly mapped to produce a knowledge that disciplines play.

Sports and Masculinity

Masculinity is a moving target. It is thoroughly historical, contextual, and relational.[30] Even setting aside the configuration of masculinity vis-à-vis femininity, different and competing notions of what it means to be a man populate the same cultural formations. This is acknowledged by the substantive body of research on hegemonic masculinity, a term that points to the cultural force of certain constructions of masculinity and the subordinate status of others. In this section I will distinguish between the terms *masculinities* and *hegemonic masculinity* in order to establish the relationship between play and masculinity.

Raewyn Connell's seminal work, *Masculinities,* which critiques psychological and sociological models of singular, static sex roles and is often credited as a foundation for the field of men's studies, is a productive starting point. Most important, Connell reminds us that there is no one masculinity but rather a range of distinct masculinities that occupies culturally specific raced, classed, and sexualized coordinates. Rural, white youths will idealize a notion of manhood that likely bears some resemblances to but is still distinct from the idea of masculinity that prevails among urban Latinos of the same age, an effect of the plethora of differences and intersections that shape cultural and communal norms. Connell further distances masculinities from biological determinations and sets of culturally intelligible types for men to inhabit when she argues that masculinity is instantiated through "body-reflexive practices," ways of being that are lived and therefore dynamic. Subjects negotiate these performances of masculinity, and they are both free and determined, which is to say that this process does not take place in a vacuum but rather occurs within historically situated moments characterized by specific formations of hegemonic masculinity. Drawn from Antonio Gramsci's Marxist theory of culture, hegemony refers to the "'spontaneous' consent given by the great masses of the population to the general direction imposed on social life." Connell's concept of hegemonic masculinity refers, then, to the notion of manliness paradigmatic of a social formation: "It was not normal in the statistical sense; only a minority of men might enact it. But it was certainly normative. It embodied the currently most honored way of being a man."[31] As the prevailing archetype of manliness, hegemonic masculinity is a focal point around which other subordinate and allied masculinities are organized. Regardless of the particular masculinity enacted through one's body-reflexive practices, hegemonic masculinity is a regulatory norm against which all masculine gender performance is measured.

In the contemporary global West, hegemonic masculinity is characterized by physical force and violence, occupational achievement, patriarchal domination of women and children, the adventurous individualism of the frontiersman, and heterosexuality.[32] While these elements are intertwined, the physical embodiment of masculinity is the keystone that binds them together. As Connell writes, "The physical sense of maleness and femaleness is central to the cultural interpretation of gender. Masculine gender is (among other things) a certain feel to the skin, certain muscular shapes and tensions, certain postures and ways of moving, certain possibilities of sex."[33] This is because "bodies are both objects of social practice and agents of social practice," which means that the body is not only worked and reworked by masculine norms but also simultaneously the vehicle that enables other performances of masculinity and a visual spectacle signifying manly capability.[34] The body was not always so vital to hegemonic figurations of masculinity, and in fact its increasing centrality can be traced in parallel with the movement from feudal to industrial social organization as aristocratic ideals of chivalry and noblesse oblige were complemented by concern for physical toughness, virility, and strength.[35] In the midst of this shift, the moral and intellectual authority of the Enlightenment, and Kantian philosophy in particular, legitimated the equation of manhood and reasoned thought.[36] But, again, the body serves as the signifier of the mind, subordinating actual intellectual competence to mastery of one's own physical form.

As Connell explains, sport is a vital forum for practicing and demonstrating this mastery. "Sport has come to be a leading definer of masculinity in mass culture. Sport provides a continuous display of men's bodies in motion. Elaborate and carefully monitored rules bring these bodies into stylized contest with one another. . . . It is the integrated performance of the whole body, the capacity to do a range of things wonderfully well, that is admired in the greatest exemplars of competitive sport."[37] Strength, vigor, vitality, and mastery over both oneself and others are demonstrated in sport. However, the contemporary fetishization of muscularity was a twentieth-century invention. Born of a slew of cultural, economic, demographic, and geographic changes that threatened white male privilege, the figure that came to exemplify masculine muscularity was Eugene Sandow. The original bodybuilder, Sandow toured North America and Europe, his body a spectacle of muscle and strength. Notably, Sandow's show was promoted as an opportunity to see the "world's best developed man," a slogan that resonated with the cultural desire to (re)constitute the power of manhood.[38] Recently, organized sports have been recentered as sites where masculinity is produced and reified as a backlash against feminism, just as the early twentieth century saw a cultural investment in sports in the wake of and response to incorporation of women into the public sphere during the Victorian era.[39] Additionally, the massive, sometimes global media dissemination of contemporary sports makes them a key site for exhibiting and reproducing hegemonic masculinity.[40]

Though pure play and gameplay are not mutually exclusive with hegemonic masculinity, it is much more coherent with sportive play. For one, occupational success is possible only with organized sports, as professionalization is contrary to pure play and a sign of the transformation of gameplay into rationalized, sportive play. Additionally, sports symbolically and materially enact patriarchy through, on the one hand, the hierarchal relations between typically male athletes and female cheerleaders and, on the other hand, the cultural and economic dominance of men's sports relative to women's sports. Play, however, militates against domination. And even though "most modern elite sport is a highly disciplined practice subject to intense surveillance . . . maverick sports stars appear to offer the power to live the life of masculine individualism, defying constraints and rebelling against regulation whilst still performing."[41] There is little opportunity in gameplay for superstars to emerge, and the ideal of a playground phenom is silly. Most important, however, sport is a well-built stage for the spectacle of the well-developed body of hegemonic masculinity. Granted, the body can be displayed in other forms of play. Both pure play and gameplay can, at times, enable feats of physical stamina, strength, and coordination. But only in sport is the sustained, practiced mastery of one's body essential to not only play but play well in a manner that enables winning. Energy must be conserved, muscles expended and relaxed to preserve continued performance, and maximum effect achieved through the exertion of the least requisite force. Despite the appearance of physical excess, the body is restrained, its performance an exemplar of instrumental rationality. Of course, sport is not the only site where masculinity is produced, and other notions of masculinity have cultural capital, too. Michael Kimmel describes the cultural moment in which hegemonic masculinity was solidifying around the muscular, sportive body as a national "crisis of masculinity" fueled by urbanization, changing modes of production, and an increasingly diverse population.[42] As traditional avenues of skilled labor simultaneously shrank and came into competition with new waves of immigrants, new technologies offered an alternative figure of manhood, as did the increasing footprint of corporate culture, which necessitated an expanded managerial class.

One notion of masculinity allied to but still very much distinct from hegemonic masculinity is the figure of the everyday Peter Pan. Woody Register highlights American entrepreneurs Tony Sarg and Oscar Hammerstein, as well as the commercially successful artists Victor Herbert, Frank Lloyd Wright, and L. Frank Baum, as everyday Peter Pans who "helped lay the institutional foundation of twentieth century consumer capitalism." Like hegemonic masculinity, the everyday Peter Pan was operative in the first several decades of the twentieth century. However, instead of reasserting the primacy of the (male) body to counteract the protofeminist gains of the Victorian era, the everyday Peter Pan harnessed the feminine "appeal of magic, fantasy, theatricality and desire" for commercial gain. By leveraging the ambivalence of the association between play, femininity, and childhood, he staged the "destabilizing, carnivalesque tendencies of the new economic world" to create a location for manhood within the coordinates of an increasingly consumer-oriented culture in which labor and production were both literally and figuratively losing currency. This play was both vigorously visceral, though undisciplined, and immersively imaginative as well. In this sense, it may very well be that the everyday Peter Pan helped eased the friction of the passage from aristocratic to bourgeois ideals of manliness while simultaneously making possible its articulation with hegemonic masculinity.[43]

Another set of constructions more directly in competition with hegemonic masculinity centers manhood on technological competence. While this notion of technomasculinity is certainly tied to the Enlightenment discourse linking manliness to reason, it is also something more to the extent that it premises manhood on the application, or external display, of reason through the making and mastering of technologies. Instead of locating the outward sign of manliness in the body, creation and use of technology become a sign of power. Arguing that the "ideology of masculinity has [an] intimate bond with technology," Judy Wajcman discusses the scientific communities responsible for the development of computers and the atomic bomb. She highlights an obsession with control over the intellect and the scientific process that is complemented with discourses of the "tough and fast," "shoot-from-the-hip" nature of the work that conjure associations with hegemonic masculinity. Carolyn Marvin's study of the emergence of electronic technologies in the late nineteenth and early twentieth centuries, and the discourses that assigned social value to them, is also instructive. She identifies a concerted effort to elevate the social position of electrical experts – scientists, engineers, and technicians – by virtue of their ability to manipulate and maintain electrical technologies in a social climate in which "ignorance regarding the use of technology was a virtue of 'good' women." However, this valorization of technological prowess still showed some deference to hegemonic masculinity, as illustrated by the dystopian discourse of the "pushbutton," which envisions a world in which technology has gone awry and men have become weak and flaccid as a result of automation.[44]

While technical expertise and entrepreneurial acumen remain viable means of demonstrating manliness, they do not typically challenge hegemonic masculinity but rather set it in relief. However, in the sportive play of digital games, these qualities come together with elements of hegemonic masculinity in an unlikely performance of manhood.

THE SPORTIFICATION OF DIGITAL GAMES

Not every game fosters competition between players, but digital games have always enabled it. Even during the era of single-player arcade games, players competed for high scores asynchronously and often across great distances, though in some instances, such as the early 1980s television show *Starcade,* players were brought together to compete head-to-head and arcade cabinet-to-cabinet.[45] In the first few years of the 1990s, "deathmatches" played over LAN networks cemented the popularity of first-person shooter games *Doom* and *Quake,* which revolutionized competitive digital gaming by shifting the scene from singular sites such as a living room or den to multiple, geographically distant sites connected through the internet. The creator of *Doom* and *Quake,* iD Software, would also host a *Quake* tournament in 1997, one of the first of its kind.

The same year saw the emergence of the Cyberathlete Professional League (CPL), currently inactive but recently acquired by a Chinese investment firm, which began with an annual tournament in Dallas, Texas, that became a biannual event the very next year. While the first tournament offered competitors prizes in the form of merchandise worth approximately $4,500, the first CPL World Tournament, held in 2001, offered $150,000 in prize money and a $40,000 purse for the winner of the main event, a one-on-one *Counter-Strike* deathmatch. From 2004 to 2007, the CPL World Tour was web-streamed from events in Chile, Italy, Singapore, and the United Kingdom on a pay-per-view basis to thousands of fans worldwide. At its height in 2005, the CPL World Tour featured a $150,000 purse for its main event and $1 million in total prize money.

A number of organizations have endeavored to mimic the early successes of the CPL, but two entities have a legitimate claim to having picked up the torch, the World Cyber Games (WCG), first held in 2000, and Major League Gaming (MLG), founded in 2002. Whereas contestants from sixteen countries were invited to compete in the inaugural WCG, in 2008 eight hundred players from seventy-eight countries competed in national qualifiers in order to play in the games in Cologne, Germany (and in 2012 players from fifty-eight countries were represented). Featuring team and individual events, the WCG envisions and markets itself as the Olympics of the digital era.[46] Between WCG events, most players competitive on the world stage play year-round in leagues. Major League Gaming, the largest North American league, has a robust community following and a declared intent to model itself on other North American professional sports league. Until the 2012 season, thousands of players competed in online qualifiers in order to advance to circuit tournaments and then to the annual championships, though a number of professional teams are awarded berths as well. In the past year, the league structure changed so that teams and individual players who perform well at semiannual championship tournaments receive bids to play in invitation-only "arena" events. MLG events, primarily championship tournaments but some circuit events too, have been broadcast on the cable networks USA and ESPN.

Though not every tournament contestant can call digital gaming their day job, for some cyberathletes gaming is not only full-time work, but a profession. The average MLG-sponsored professional player makes an annual salary of between twenty and thirty thousand dollars, while exceptional, top-ranked players make up to eighty thousand dollars per year of salary, plus any prize money won and endorsement revenue.[47] Gaming is certainly, for some, much more than a pastime; it is a profession.

PATTERNS OF PROFESSIONAL GAMEPLAY

Not everyone finds the institutionalization, monetization, and broadcasting of competitive digital gaming as warranting the title of "e-sports." A prominent refrain from those who question the sportiness of e-sports is the charge that gaming is not a physical activity. Another (related) assumption that troubles e-sports' claim to the status of sport is the negative image of gaming as an unhealthy and antisocial activity.[48]

I maintain that digital gaming is a corporal practice and, despite a bad reputation, helps produce players more capable of successfully navigating the contemporary social terrain. Evidence of this can be seen in the patterns of practice constitutive of sportive digital gameplay, which the works of T. L. Taylor and Nick Taylor take great pains to identify and explain in detail. I outline, here, several notable sets of practices that Taylor and Taylor argue are characteristic of e-sport.

Nick Taylor highlights two distinct communication practices germane to professional first-person shooter gaming, callouts and trash talk. Callouts consist of loud, descriptive statements aimed at one's own teammates. They tell teammates where a player or opposing players are, how much damage enemies have taken or how much it will take to eliminate them, and the status of items or locations in the game arena. Callouts are highly stylized, and the very terms employed in callouts are standardized into what Taylor terms "a highly codified terminology referring to specific areas in specific game arenas." Players cannot expect to invent descriptive statements and expect to meaningfully communicate with teammates but must, rather, visit game-related discussion forums to study diagrams and memorize abbreviations.[49] Even players who do not speak English as their primary language employ the standard English-language terms.[50] The use (and nonuse) of callouts is a strong indicator of a player's relationship, as an insider or outsider, to competitive first-person shooter gaming and is policed by skilled players.[51] Typically neither imperative nor directive, callouts nevertheless regulate gameplay by providing information that gives their teammates more information about the game's state than could be discerned alone and thus direct action.

Trash talk and taunting, while also typically loud, are aimed at opposing teams and often contain as many directive statements as descriptive. Because teams are typically seated within ten feet of each other, there is ample opportunity for rival players to insult one another's play by describing it in misogynistic or homophobic terms, question their abilities, call attention to their standing in the match or tournament, and question the wisdom or effectivity of a strategy or performance. Trash talking is aimed at disrupting the otherwise tightly choreographed and coordinated teamwork that is enabled by callouts (and extensive practice). While trash talk may seem decidedly unprofessional, it is, in fact, a typical element in competitive sports and in the MLG (but not the WCG, which aspires to Olympian standards) is considered to be a legitimate, tactical element of competitive team gameplay. As such, norms and expectations have developed around the practice so that it aims to distract and tease rather than provoke and stays on the playing field, ending with the conclusion of a match, at which point a general collegiality is reestablished with handshakes and congratulations or consolations.[52]

In addition to these communicative norms, T. L. Taylor discusses a number of routine practices, both physical and mental, and knowledges typical of the professional players competing at the MLG and WCG level. Initially, Taylor focuses on the importance of embodied skills. Digital gameplay is undeniably an embodied activity, but, nevertheless, e-sports occupy a similar ground as other liminal cases of sport, such as NASCAR, which requires "body wrenching control over a piece of technology," and darts and archery, which demand precise control over small physical movements.[53] In these sports, the quantity of physical exertion is not what counts so much as the quality of embodied practice. Even though certain game genres may require more or less physical exertion, the first-person shooters, fighting, rhythm, and real-time strategy games played at MLG and WCG events demand quick reflexes and rapid finger and hand movements.[54] This leads some to argue that e-sports are identified by the "bodily competence" of "rapid and accurate coordination between the hand and eye."[55] Taylor locates the physicality of e-sports more deeply in and diffusely throughout the body; she argues that the performance required of cyberathletes "inscribes itself on the body of players, refining over time the most nuanced yet complex circuits of action."[56] This is one of the reasons cyberathletes practice daily. Like any feat of dexterity, skilled gameplay relies as much upon muscle memory as it does conscious action, and practice is vital to training the body to respond with speed and precision beyond the capabilities of an unpracticed amateur. A range of sports, from NASCAR to darts, belie the belief that physical power is essential to sport but nevertheless rest upon a foundation of highly trained and precisely executed physical activity. Nothing captures this idea more clearly than the statistics kept by the MLG regarding professional players' "actions per minute." This statistic measures not only the speed and reflexes of players but also, because of the level at which they play, the skill inherent in quick play.

A crucial knowledge that Taylor ascribes to professional gamers is mastery of the game and game technology. This knowledge impacts embodied play in some very concrete ways, mediating a skilled player's relationship to the material configuration of the technological apparatus in the competition space. As Taylor explains, the often tight, cramped play spaces "generate a bodily discipline and self-regulation among participants" responsive to both the unwillingness to touch one another prevalent among a mostly heteronormative group of players and the necessity to play in a fluid and responsive manner. Technical knowledge also informs embodied play in unquantifiable ways, from inculcating a general "ease around technological objects" to the ability to "speak assuredly (and thus with legitimacy) about technical matters within a gamer subculture." Part of the display of technical mastery consists of players bringing their own equipment to tournaments, which typically allow outside keyboards, mice, and, more infrequently, TV screens. Mastery of the game also regulates gameplay, though its impact is upon the employment of strategy and performance of action in game. As Taylor writes, "Knowing a map, what various characters or classes can do, and weapons, as well as understanding the physics of a system and knowing basics like commands, macros, and shortcuts all form a part of game mastery."[57]

Taylor's ethnography also accounts for a second assemblage of knowledge that is characterized by tactical thinking and skilled improvisation. By virtue of their practice regimens and long, studied engagement with the games played on the competition circuit, professional players know how to react to situations in ways that maximize their chances, however slim, of winning. As Taylor describes, players run scrimmage matches not only to refine existing skills but also to develop new skills. In these practice games, different possible responses (or counters) to effective strategies employed by the professional community, or even specific teams, are explored in order to research their effectivity.[58] Elsewhere, in the context of deeply committed MMORPG players, Taylor discusses this in terms of instrumental play.[59] Unlike the engaged but informally examined gameplay of a casual player, instrumental play is rationalized to achieve the most efficient return upon an action and the product of a great deal of study of game footage, databases containing game data, and discussion forums. With experience and a mental repertoire of tactical responses to recurrent situations, highly skilled professional gamers are also able to react to novel circumstances. Taylor claims, "Elite play is interwoven with opening moves, set and known tactics, and a high dose of improvisation" that consists of employing learned skills to "maneuver dynamically based on the actual play situation at hand."[60] However, where Taylor describes this improvisational play as imaginative, it is worth noting that despite the often impromptu character of the performance, its elements are nevertheless very practiced.

NEOLIBERAL MASCULINITY

The practices and knowledges constitutive of professional gaming are telling. They not only define the contours of a nascent cultural sphere, but also replicate important aspects of the broader cultural formations from which they emerge. Professional gamers perform their identity by disregarding key aspects of hegemonic masculinity and adopting elements of unconventional masculinities. This is motivated not by a culture of tolerance, but rather by the imperative of the neoliberal cost-benefit analysis players use to determine the most effective means to their ends. This neoliberal masculinity, however, is just one way in which the neoliberal rationality that underwrites professional, sportive gaming is instantiated. At the core of this still embryonic culture is a thorough rationalization (and therefore erasure) of play that transforms each act of play into an economic calculus, normalizing the neoliberal logic upon which it is premised.

Michael Wagner proposes that "'sport' is a cultural field of activities in which people voluntarily engage with other people with the conscious intention of developing and training abilities of cultural importance as well as of comparing themselves with other people in these abilities according to deliberately accepted rules."[61] This definition clashes with Guttmann's insistence that sport is a physical contest, but resonates with his claim that, in the modern era, sport is secular, quantified, bureaucratized, specialized, and rationalized, as these qualities also define the social world. It should not, then, be surprising that contemporary e-sports train players to think and act like *Homo economicus,* basing every action upon a calculation of cost for benefit, aimed at maximizing the probability of and rewards associated with an outcome.

This can be seen in the typical performance of masculine gender identity in professional gaming. Both Nick Taylor and T. L. Taylor give careful attention to the construction and performance of masculinity in e-sport. Nick Taylor shows that competitive gaming employs hypermasculine discourses that glorify cyberathletes and high-stakes competition as a strategy to legitimate its claim to sport.[62] His analysis uncovers a fairly coherent performance of hegemonic masculinity in e-sport that is exhibited through the communicative policing of gendered space,[63] avoidance of physical contact between males,[64] and institutional marginalization of women.[65] T. L. Taylor's examination of gender in professional gaming identifies it as a site of struggle. She points to conflicting efforts to align e-sport with hegemonic masculinity and a "geek masculinity" organized around technological competence to argue that "what is happening in the pro scene, and game culture more generally, is a struggle over the status of masculinity."[66] However, these claims are problematic; where Nick Taylor accounts for only masculine performance, T. L. Taylor mistakes players' claims to identify with one or another notion of masculinity as evidence that substantive differences are enacted by players.

I argue, rather, that the performance of sportive gameplay embraces aspects of hegemonic masculinity, technomasculinity, and the playful masculinity of the everyday Peter Pan. Professional gamers train their bodies to play, but their bodily discipline is not aimed at musculature. Eschewing this cornerstone of hegemonic masculinity, they instead develop a very specific set of coordinated reflexes, localized in the hands and fingers. Still, the core of professional gaming is the competition and the opportunity to dominate an opponent through virtual violence. This focus on power is supported, in part, by the vigorous exchange of trash talk and taunts designed to disrupt and subordinate rivals. These more traditionally masculine acts are complemented by qualities associated with technomasculinity. Sportive gameplay is a flux of called plays and tactically sound practiced improvisation. As such, it requires that skilled players dedicate to memory an encyclopedic knowledge of tactics obtained through careful study of the game and verified through repeated testing. This, in turn, is enabled by an extensive knowledge of the game and the technology on which it is played. No practice better exemplifies this combination of qualities than the callouts players provide team-mates. Loud, descriptive statements about the game state, callouts aim to map out the gamespace as a means of dominating it and require comprehensive knowledge and mastery of both the game and game-specific language in order to ensure effective communication. Finally, the serious work of competitive gaming is fun (for most cyberathletes). By making their living through their leisure, professional gamers perform what Derek Burrill calls "digital boyhood . . . in which adult males return to their adolescence to play without the responsibilities of adulthood."[67] Regardless of whether they identify as an athlete or a geek, a professional gamer is in fact both, and a bit childish, too.

This is not a clash of rival masculinities but a hybridization that represents the most economic combination of skills and abilities needed to succeed in the arena of professional gaming. Indeed, the professional gamer is an exemplar "entrepreneur of the self," a subject who takes on whatever traits a cost-benefit analysis determines will best allow him or her to sell his or her labor.[68] And so the masculinity of the professional gamer is principled in only one regard; it is a neoliberal masculinity.

The logic guiding this performance of masculine identity saturates e-sports, which train players to employ an economic calculus. Gender identity is the site of application examined in this chapter, but by no means the limit of the constitutive effect of sportive play. As Wagner argues, e-sports train players in "cyberfitness," a set of competencies that he claims, turning to management theory, "optimizes human skills for maximum performance within a fixed software environment."[69] In contemporary neoliberal society, this sort of job training is ideal, and it is undertaken by far more than the forty professionals who make a living in e-sports. The thousands who play in MLG online qualifiers and the millions more who, caught up in the cultural moment, have embraced a sportive style of digital gameplay undertake the same training. They learn, by playing, to be *Homo economicus*. They practice neoliberal subjectivity in games and refine their abilities to weigh options and make decisions to minimize cost and maximize reward.

The curious reception of e-sports in China provides a final illustration. Online role-playing games that emphasize a long, slow process of character development and cooperation with other players to overcome computer-generated obstacles are looked down upon and derided in the Chinese press. However, the competitive games that are often used for e-sports, real-time strategy and first-person shooter games, are typically praised as "athletic" alternatives to their dangerous, "addictive" counterparts. In fact, e-sports are officially recognized as a professional competitive sport by the Chinese government, a move that Marcella Szablewicz links to China's economic liberalization efforts. She writes about the state-sponsored push to encourage entrepreneurship of the self, explaining that "emphasis has shifted from centrally directed 'self-sacrifice' for the nation, to autonomously choosing 'self-development' for the nation."[70] In this context, e-sports are an excellent platform for training players to apply neoliberal rationality not only to money matters, but ubiquitously, to cultivate the self and reconstitute society.

NOTES

1. Allen Guttmann, *From Ritual to Record: The Nature of Modern Sports* (New York: Columbia University Press, 1978), 11.

2. Wendy Brown, *Edgework: Critical Essays on Knowledge and Politics* (Princeton, NJ: Princeton University Press, 2005), 37.

3. Jason Read, "A Genealogy of *Homo-Economicus*: Neoliberalism and the Production of Subjectivity," *Foucault Studies 6* (2009): 28.

4. Thomas Lemke, "'The Birth of Bio-Politics': Michel Foucault's Lecture at the Collège de France on Neo-liberal Governmentality," *Economy and Society* 30, no. 2 (2001): 197.

5. Michel Foucault, "Governmentality," in *The Foucault Effect: Studies in Governmentality,* edited by Graham Burchell, Colin Gordon, and Peter Miller (Chicago: University of Chicago Press, 1991), 88, 102.

6. Bruce Curtis, "Foucault on Governmentality and Population: The Impossible Discovery," *Canadian Journal of Sociology* 27, no. 4 (2002).

7. Michel Foucault, "Security, Territory, and Population," in *The Essential Works of Foucault, 1954–1984,* edited by Paul Rabinow (New York: New Press, 1997), 1:70.

8. Foucault, "Governmentality," 92, 99.

9. Michel Foucault, *Security, Territory, Population: Lectures at the Collège de France, 1977–1978,* edited by Michael Senellart and translated by Graham Burchell (New York: Palgrave Macmillan, 2009), 59.

10. Tony Bennett, "Culture and Governmentality," in *Foucault, Cultural Studies, and Governmentality,* edited by Jack Bratich, Jeremy Packer, and Cameron

McCarthy (Albany: State University of New York Press, 2003), 60.

11. Barbara Biesecker, "Michel Foucault and the Question of Rhetoric," *Philosophy and Rhetoric* 2 (1995): 354.

12. Michel Foucault, "Technologies of the Self," in *Technologies of the Self: A Seminar with Michel Foucault,* edited by Martin Luther, Huck Guttman, and Patrick Hutton (Amherst: University of Massachusetts Press, 1988), 19.

13. Trent H. Hamann, "Neoliberalism, Governmentality, and Ethics," *Foucault Studies* 6 (2009): 38.

14. Lemke, "'Birth of Bio-Politics,'" 199.

15. Randy Martin and Toby Miller, "Fielding Sports: A Preface to Politics?," in *SportCult,* edited by Randy Martin and Toby Miller (Minneapolis: University of Minnesota Press, 1999), 10.

16. Brian Sutton-Smith, *The Ambiguity of Play* (Cambridge, MA: Harvard University Press, 1997), 1–2.

17. Robert Ancho, "History and Play: Johan Huizinga and His Critics," *History and Theory* 17, no. 1 (1978): 63.

18. Johan Huizinga, *Homo Ludens: A Study of the Play Element in Culture* (Boston: Beacon Press, 1955), 4, 9; Jean Piaget, *Play, Dreams, and Imitation in Childhood* (New York: W. W. Norton, 1962), 147; Huizinga, *Homo Ludens,* 10.

19. Roger Caillois, *Man, Play, and Games* (New York: Free Press, 1961), 13.

20. Immanuel Kant, *Critique of Judgment,* translated by J. H. Bernard (London: Macmillan, 1914), 59.

21. Hilde Hein, "Play as an Aesthetic Concept," *Journal of Aesthetics and Art Criticism* 27, no. 1 (1966): 67.

22. Piaget, *Play, Dreams, and Imitation,* 85.

23. Caillois, *Man, Play, and Games,* 26, 9; Piaget, *Play, Dreams, and Imitation,* 168.

24. Susan Bordo, "Feminism, Postmodernism, and Gender Skepticism," in *Unbearable Weight: Feminism, Western Culture, and the Body* (Berkeley: University of California Press, 2003), 226.

25. Jacques Derrida, "Structure, Sign, and Play in the Discourse of the Human Sciences," in *Writing and Difference,* translated by Alan Bass (London: Routledge, 1978), 279.

26. Foucault, "Technologies of the Self," 17–18.

27. Michel Foucault, "The Subject and Power," in *Power: The Essential Works of Michel Foucault, 1954–1984,* edited by J. D. Fabion and translated by R. Hurley (New York: New Press, 2000), 3:342.

28. Caillois, *Man, Play, and Games,* 15; Guttmann, *From Ritual to Record,* 14.

29. Guttmann, *From Ritual to Record,* 7, 85, 89; Toby Miller, "Competing Allegories," in *SportCult,* edited by Martin and Miller, 18.

30. Raewyn Connell, *Masculinities* (Cambridge: Polity Press, 1995), 43–44.

31. Connell, *Masculinities,* 76, 57; Antonio Gramsci, "The Formation of Intellectuals," in *Selections from the Prison Notebooks* (London: Electric Book, 1999), 145; Raewyn Connell and James Messerschmidt, "Hegemonic Masculinity: Rethinking the Concept," *Gender and Society* 19, no. 6 (2005): 832.

32. Nick Trujillo, "Hegemonic Masculinity on the Mound: Media Representations of Nolan Ryan and American Sports Culture," *Critical Studies in Mass Communication* 8 (1991): 291–292.

33. Connell, *Masculinities,* 52–53.

34. Connell and Messerschmidt, "Hegemonic Masculinity," 851.

35. Alan Peterson, *Unmasking the Masculine: Men and Identity in a Skeptical Age* (London: Sage, 1998), 42–43.

36. Victor Seidler, *Rediscovering Masculinity: Reason, Language, and Sexuality* (New York: Routledge, 1989), 2.

· 37. Connell, *Masculinities*, 54.

38. Jon Stratton, "Building a Better Body: Male Bodybuilding, Spectacle, and Consumption," in *SportCult,* edited by Martin and Miller, 156–157.

39. Peterson, *Unmasking the Masculine,* 47; Connell, *Masculinities,* 54.

40. Trujillo, "Hegemonic Masculinity on the Mound," 292.

41. Cara Aitchinson, "Gender, Sports, and Identity: Introducing Discourses of Masculinity, Femininity, and Sexuality," in *Sport and Gender Identities: Masculinities, Femininities, and Sexualities,* edited by Cara Aitchinson (New York: Routledge, 2007), 14–15.

42. Michael Kimmel, "Baseball and the Reconstruction of American Masculinity, 1880–1920," in *Sport, Men, and the Gender Order: Critical Feminist Perspectives,* edited by Michael A. Messner and Donald F. Sabo (Champaign, IL: Human Kinetics Books, 1990), 158.

43. Woody Register, "Everyday Peter Pans: Work, Manhood, and Consumption in Urban America, 1900–1930," in *Boys and Their Toys? Masculinity, Technology, and Class in America,* edited by Roger Horowitz (New York: Routledge, 2001), 200, 204, 205.

44. Judy Wajcman, *Feminism Confronts Technology* (University Park: University of Pennsylvania Press, 1991), 137, 141; Carolyn Marvin, *When Old Technologies Were New: Thinking about Electric Communication in the Late Nineteenth Century* (New York: Oxford University Press, 1988), 23, 124.

45. T. L. Taylor, *Raising the Stakes: E-Sports and the Professionalization of Computer Gaming* (Cambridge: MIT Press, 2012), 3–4.

46. Brett Hutchins, "Signs of Metachange in Second Modernity: The

Growth of E-Sport and the World Cyber Games," *New Media and Society* 10, no. 6 (2008): 858.

47. Associated Press, "*Guitar Hero* Whiz Aiming Higher," *Education Week,* August 4, 2008, http://www.edweek.org /ew/articles/2008/08/04/140598ncex changeguitarhero_ap.html?qs=guitar +hero.

48. Kalle Jonasson and Jesper Thiborg, "Electronic Sport and Its Impact on Future Sport," *Sport in Society* 13, no. 2 · (2010): 287.

49. Nick Taylor, "'A Silent Team Is a Dead Team': Communicative Norms in Competitive FPS Play," in *Guns, Grenades, and Grunts: The First-Person Shooter,* edited by Gerald Voorhees, Joshua Call, and Katie Whitlock (New York: Continuum, 2012), 258, 259.

50. Nick Taylor, "Play Globally, Act Locally: The Standardization of Pro *Halo* 3 Gaming," *International Journal of Gender, Science, and Technology* 3, no. 1 (2011): 234.

51. Taylor, "Silent Team," 262–263, 267–268.

52. Ibid., 267, 266.

53. Emma Witkowski, "Probing the Sportiness of eSports," in *eSports Yearbook, 2009,* edited by Julia Christophers and Tobias Scholz (Norderstedt, Germany: Books on Demand, 2009), 55.

54. Gerald Voorhees, "I Play Therefore I Am: Sid Meier's Civilization, Turn-Based Strategy Games, and the *Cogito,*" *Games and Culture* 4, no. 3 (2009): 263.

55. Jonasson and Thiborg, "Electronic Sport and Its Impact on Future Sport," 290.

56. Taylor, *Raising the Stakes,* 39.

57. Ibid., 92, 93.

58. Ibid., 94.   ·

59. T. L. Taylor, *Play between Worlds: Exploring Online Game Culture* (Cambridge: MIT Press, 2007), 74.

60. Taylor, *Raising the Stakes,* 94.

61. Michael Wagner, "Competing in Metagame Space: Esport as the First Professionalized Computer Metagames," in *Space Time Play: Computer Games, Architecture, and Urbanism, the Next Level*, edited by Friedrich von Bories, Seffen P. Waltz, and Matthias Böttger (Berlin: Birkhauser Verlag, 2007), 182.

62. Taylor, "Play Globally," 230; Nick Taylor, "Cheerleaders/Booth Babe/*Halo* Hoes: Pro-gaming, Gender, and Jobs for the Boys," *Digital Creativity* 20, no. 4 (2009): 240.

63. Taylor, "Silent Team," 268–270.

64. Taylor, "Play Globally," 236–237.

65. Taylor, "Cheerleaders," 240.

66. Taylor, *Raising the Stakes*, 117.

67. Derek Burrill, *Die Tryin': Videogames, Masculinity, Culture* (New York: Peter Lang, 2008), 15.

68. Michel Foucault, *The Birth of Biopolitics: Lectures at the Collège de France, 1978–1979*, edited by Michael Senellart and translated by Graham Burchell (London: Picador, 2010), 226.

69. Wagner, "Competing in Metagame Space," 183.

70. Marcella Szablewicz, "From Addicts to Athletes: Participation in the Discursive Construction of Digital Games in Urban China," in *Selected Papers of Internet Research* (October 2011), http://spir.aoir.org/index.php/spir/article/view/35, 10, 14.

FOUR

The Social and Gender in Fantasy Sports Leagues

Luke Howie and Perri Campbell

INTRODUCTION: THE RIGHT KIND OF EYES

Since the mid-1990s, fantasy sports participation has grown at a significant rate. According to the Fantasy Sports Trade Association, just under thirty-two million people over twelve years old play fantasy sports in North America (including Canada). In the United Kingdom, two million people participate in fantasy Premier League soccer games.[1] The financial impact of fantasy sports is measured in billions of euros and dollars.[2] Most people cite early 1980s rotisserie baseball leagues as the precursors to the contemporary online fantasy sports experience, but there is some evidence that rudimentary forms of fantasy sports have existed since the mid-1950s.[3]

In this chapter we provide an account of ongoing research being conducted with the members of a long-running fantasy NBA league based in Australia and their wives and partners. The league began in the late 1990s as a hobby for ten friends who had attended high school or college together, and all played competitive basketball, some to professional and semiprofessional levels. It is played online but features many offline supplementary activities, including a live offline draft party, elaborate mechanisms for choosing the draft order (referred to by participants as the "lottery"), and detailed, ongoing discussions about strategies, statistics, and trades taking place year-round. There are benefits to understanding fantasy sports as a site for advertising and marketing, as gambling, and as a problematic regulatory field.[4] We are studying fantasy sports leagues as a social occurrence that takes place in online and offline realms where gender matters and takes various hegemonic forms.

We believe that understanding fantasy sports involves cultivating what Hunter S. Thompson described as the "right kind of eyes." With the right eyes, we can see things in particular ways, perhaps in ways that were not immediately apparent at first glance. We see Thompson's metaphor as a methodological lesson, one that informs our use of narrative analysis methods and techniques. This type of approach can be seen in Catherine Kohler Riessman's and Donna Haraway's analyses of witnessing social events and storytelling.[5] Central to Riessman's and Haraway's approach is the task of *locating* the researchers. When we do this, we find ourselves to be insiders in fantasy sports leagues, one as a fantasy sports participant and fifteen-year veteran of Yahoo! Fantasy NBA leagues, the other as the girlfriend, partner, and now wife of a fantasy sports enthusiast.

This chapter features stories that account for a practice that is both "intensely corporeal" and "kinetic" and played out in and between online and offline social spaces.[6] We are insiders in fantasy sports leagues. One of us has played Yahoo! Fantasy NBA his entire adult life, has dabbled in other fantasy sports, and is a sporting fanatic. His Australian summers are filled with basketball statistics and 5:00 AM alarms to watch college football games. The other is the wife of a fantasy sports enthusiast. As a Muay Thai enthusiast, sport plays a central place in her social and married life, but one that is very different from her husband's. Vacations to America must take into consideration when the Lakers play at the Staples Center, and Christmas gifts are purchased on NBA.com and the popular sporting apparel website Eastbay. For Australians, it is a form of life made possible by a late-1990s explosion in the use of internet technologies.[7]

In this chapter we analyze data collected in interviews conducted with fantasy NBA enthusiasts and the wives of fantasy sports enthusiasts. If this and what follows appears as a heteronormative account, it is because that is how this fantasy sports league is organized – in many ways men act as men, and women appear to support their endeavors. In this way, the respondents in this research fit within the demographic conditions of fantasy sports league participation made up mostly of educated, professional, straight men in their thirties.[8] Featured in this chapter are the stories of six people – three enthusiasts and three wives of enthusiasts. To contextualize the data analysis, we provide a narrative of the social and gendered aspects of sport and fantasy sports leagues by drawing on the literature in these fields.

As Davis and Duncan have argued, gender relations in fantasy sports matter.[9] Sporting participation and spectatorship have long been considered a domain for the practice of hegemonic masculinity. While fantasy sports leagues undoubtedly retain some of these features, our data and analysis show that women play a significant role in the participation behaviors of fantasy sports enthusiasts. Their involvement can often be ambivalent and problematic and may also reproduce certain heteronormative practices and reinforce what it means to be hegemonically male.

FANTASY SPORTS, THE BODY, AND GENDER

Much of the early theorizing of the social aspects of the internet assumed that encounters in cyberspace were somehow fraudulent, fake, inauthentic, and *disembodied*. As the trope goes, on the internet no one knew you were a dog. This was a theme in Turkle's early work on the social features of the internet and Kitchin's work toward "geographies of cyberspace." Kitchin argued that the most prominent theories of meaning making concerning "cyberspace and identity" revolved around ideas of disembodiment or the human-machine hybrid made popular through Donna Haraway's timely identity theory of *cyborgs*.[10] In each sense, the body became a site of anxiety for understanding online social worlds.

Gender has long been on the sport research agenda. Messner et al. argued that "the institution of sport has provided men with a homosocial sphere of life through which they have bolstered the ideology of male superiority." By excluding women altogether, or by making women a relational feature of sporting participation and fandom, sport has become associated with the supposedly masculine ideals of "physical competence, strength, power, and even violence." As such, sporting participation and fandom are spaces where hegemonic masculinities are reconstituted as naturalized superiority. As Toby Miller has demonstrated, the role of women in sport has often been a site for gendered debates, ambivalence, and stinging bias and discrimination, perpetuating myths of biological male dominance.[11]

R. W. Connell, perhaps the most important scholar of masculinities, argues that "sport has come to be the leading definer of masculinity in mass culture." How should we interpret the role of the body, Connell asks, when it refuses to "stay outdoors in the realm of nature and reappear[s] uninvited in the realm of the social?" Bodies should not been viewed as biologically stable or neutral. What we need, following Turner, is an engagement with the body on different terms (even when it seems absent).[12] When we return bodies to social realms on the internet, we return gender in the process. This is particularly important, given the ways in which fantasy sports research is growing. In much research, the role of the participant, the enthusiast, the player, and their bodies has been reduced to numbers and text. We want to return the body to fantasy sports and, in the process, frame fantasy sports leagues as socialized and gendered spaces that blend online and offline sporting fandom. When media and sport come together, we often end up with an "overwhelmingly male and hegemonically masculine domain that produces coverage by men, for men and about men." In this sense, fantasy sports can often be a place that "valorizes elite, able-bodied, heterosexual, and professional sportsmen, especially those who bring glory to the nation."[13]

Our study owes much to a similar study by Nikolas W. Davis and Margaret Carlisle Duncan. They argue that "sociological studies have frequently presented the sport domain as a major site for reinforcing hegemonic masculinity by creating and recreating what it means to be a man through masculine interaction." Indeed, this was also among our interests in our first publication from this research project, which began formally about two years ago and will continue into the foreseeable future.[14] As fantasy NBA insiders and observers, we have been living this project since about 1999. In that paper we examined the message boards in the same long-running Yahoo! Fantasy NBA league and witnessed some of the dimensions and displays of hegemonic masculine behavior described by Davis and Duncan, especially in relation to the "harsh, sexist vernacular used by participants."[15]

We differ with these researchers, however, in their claim that there is a "relative paucity of female participation" in fantasy sports.[16] In one sense, this statement is broadly true. Women generally do not manage fantasy sports teams.[17] But women are central to the conduct and functioning of fantasy sports leagues. Women are significant figures in male team sports as both a "bodily reality" and a "category of analysis."[18] Some women position themselves in opposition to the fantasy sports to which their partners and husbands are devoted. This was the case for the organization known as WAFS, Wives against Fantasy Sports, which was a support group for "widows" during sports seasons (which, if you're doing it "right," is year-round).[19] Other women, like the women in this study, are supporters or perhaps even enablers. They recognize that fantasy sports are loved by the people they love, and the fantasy sports women featured in this chapter have taken steps to be included in the game. Where once this time spent on fantasy sports may have been considered *man's time,* it is now a space where manly men engage in masculinized sporting fandom alongside significant others who – in some case – become their number-one fantasy fans. The grief associated with injuries, contract disputes, and unexpected loss of form is shared, as is the consumer experience of buying viewing packages, sports jerseys, and caps to enhance the fantasy sports experience. The first author's wife consoled him when Rajon Rondo's season was prematurely ended by a serious knee injury. When he learned of the injury, he was wearing the Boston Celtics shorts she had bought him for Christmas from Eastbay. Indeed, we believe that the appearance of a lack in women's participation can be better explained by the general paucity of empirical research into fantasy sports enthusiasts.[20]

Recent research has advanced our understanding of the role of fantasy sports in everyday life substantially. Lee et al.'s methodologically sophisticated study (involving large-scale quantitative methods and analysis) has provided an indication of women's in-game participation based on a sample of 283 respondents drawn from two online message boards where fantasy sports discussions occur (one devoted to fantasy football, the other to fantasy baseball).[21] Four percent of respondents were women. The study also provided other demographic data. Eighty-nine percent of respondents were white, fifty-six percent had a college degree, and forty-one percent reported an annual income of at least fifty thousand dollars. This is in line with other research and provides compelling evidence that fantasy sports participation remains a pastime primarily for financially secure white men. As we have elsewhere documented, the participants in the league studied here meet this demographic picture.[22] Farquhar and Meeds's study provides insight into the motivations for participation among 42 fantasy sports participants (38 men, 4 women). They suggest that "social interaction" plays much less of a role in fantasy sports leagues than previously thought.[23]

We are intrigued by this conclusion, since many of the motivating factors described in their study as nonsocial are better explained by social factors discovered in our research. The principal motivating factors in Farquhar and Meeds's research are *arousal* and *surveillance*. Arousal refers to the "thrill of victory," and surveillance refers to the reduction of complex sporting outcomes into statistical realities that can be compared and analyzed and that form the basis of in-game decision making. These forces can easily be interpreted as social factors when this study is situated in the broader literature. We believe that "bragging rights," a factor identified by Lee et al., might be a mechanism through which the thrill of victory combines with statistical surveillance in the pursuit of bettering one's opponent. We argue that "bragging rights" are a strongly social factor. Lee et al. also downplay the social elements of fantasy sports and contextualize participation in disembodied terms, where fantasy sports provide participants with the ability to live "a vicarious life."[24] We intend to demonstrate that fantasy sports are an important part of the social lives of the fantasy sports enthusiasts and the wives and partners of fantasy sports enthusiasts whom we interviewed. By doing so, we hope to offer an alternative narrative, showing that many of the factors identified in the quantitative analysis conducted by Farquhar and Meeds and Lee et al. are more social than they first appeared.

The fantasy league we studied began with a group of like-minded friends, some of whom attended high school together and others who attended college together. They were eighteen to twenty-one years old when they began their league, and they are all now in their early to midthirties. All are straight men, all bar one are in long-term relationships or married, two have children, and they all work in professional industries. Most are white. While their league is hosted on the Yahoo! Fantasy NBA platform, most of their enjoyment comes from the *real* events that they hold to facilitate the smooth functioning of their league. The main event in the fantasy NBA calendar is the live offline fantasy draft.[25] Participants also hold a live event to determine the draft order. In July 2013 the method for selecting the draft order involved eight out of ten league members congregating at a local bar to watch an Australian Rules Football game. Each manager selected what he believed the final margin would be, with closest to the real margin winning the first pick and furthest from the margin winning the last pick (in descending order).[26] Six people from this league have been interviewed for this chapter – three are male fantasy sports enthusiasts, and three are women, all wives of fantasy sports enthusiasts. The interviews were semistructured and open-ended, and free discussion was encouraged. Three distinct issues emerged from the interview process: participating in fantasy sports is a social experience, people are highly active in fantasy sports while at work, and fantasy sports provide a way for gender to be *performed,* sometimes in ways that reinforce gendered norms and hegemonic masculinities.

FANTASY SPORTS AS A SOCIAL ACTIVITY

We argue that socializing is central to the conduct and functioning of fantasy sports. The fantasy sports gamers who were interviewed for this research project are all offline friends. They have all been dedicated and skilled basketballers, most of whom still play in club and social competitions. Fantasy sports happened in inherently social situations for these enthusiasts and members of a long-running fantasy NBA league. Their lives as friends had become enmeshed with fantasy sports experiences, and, as Davis and Duncan similarly argued, fantasy sports regularly "bleeds into everyday conversation and leisure time." This became clear to Al, an academic and psychologist in his midthirties who is married to Chanel, when he watched NBA games on television: "I get to watch a game somewhere, and that is more of a social thing. It is fantasy because I sit there barracking for my player like a maniac, for whoever I've got on my team. . . . It's more about sitting there watching basketball with my mates and barracking for that one specific player. It's less fantasy than when I am *not* watching basketball."[27]

Many of the league participants had been friends since high school and college, but Al believed that their shared love of fantasy NBA meant that some friendships have endured as a result of the bonds forged by participating in the same fantasy sports league: "There are probably four guys in this league that I am sure I wouldn't talk to anymore. Not because I don't like them, but if I hadn't been playing basketball with them for the last twelve years [on reflection he realized it is closer to fourteen years] . . . I would probably only regularly talk to three or four people from the league, and those guys not nearly as much." So whereas for Chanel, a psychologist in her early thirties and Al's wife, "It's like my husband has the hobby of a ten-year-old," she acknowledged its importance in her husband's life: "It keeps him engaged with all his mates."[28] Indeed, because the only risk in playing fantasy sports was, according to Al, wasting time, the social benefits alone justify continued participation. Hakeem, a married engineer in his midthirties, explained how fantasy sports became a facilitator for ongoing social activities and friendships built around a mutual love of sports:

> Early on it may not have been [about socializing and friendship] because I hardly knew [the] guys, although we were at Uni [college] and that kind of stuff. But now, it's one of the important ones [motivations for playing]. It's fun. I love draft day *so much*. . . . Just getting together with [the] guys. . . . I don't really see [the] guys as much as the other guys see each other. . . . I *love* it. That's why before the draft I'm nervous; you won't get a word out of me. I'm focused. Afterwards, give me a drink, let me relax, let me absorb what just happened . . . Let the fun begin! Before then it is no fun for me.[29]

For Hakeem, the fun began when the socializing started. The members of this league have a large party each year after they draft their teams via a live pen-and-paper draft, which later is uploaded onto the online platform. These parties had sometimes continued into the next day and often involved heavy drinking and a near-constant stream of discussion about the coming fantasy season. Perhaps the close, preexisting social ties that bound the members of this league explain why our findings dif-. fer from other studies that have concluded that the social elements are not so important in fantasy sports participation.[30] In the league studied for this project, there was no fantasy league without the social interactions that the league supported. Carlos, a married sales manager in his midthirties, spoke passionately about how important fantasy NBA was for bonding with his closest friends: "It enhances social situations. It enhances the relationship I have with existing friends. . . . [T]he extension of that is building on friendships . . . maintain[ing] that relationship [when] otherwise I wouldn't even hang out with them. It's camaraderie. Been together in the league for 'X' amount of years. It's quite unique to be able to have a group that meets on an annual basis that continues to be in touch with one another through this particular avenue."[31]

It is here that some critical reflection on Lee et al.'s and Farquhar and Meeds's studies is required. Given the way that Lee et al. recruited participants for their studies (via online fantasy sports message boards), it may be that those who frequent discussion boards and sites are trying to add the social element into their experiences of fantasy sports. Farquhar and Meeds recruited from a group of college students. If the researchers conducted this research again today, they might find that the leagues they studied had persisted and strong social elements had emerged. We suspect that further research, perhaps involving mixed methods, will reveal that social factors are the primary motivators for fantasy sports participation. Indeed, if we reimagine the factors identified by Lee et al. and Farquhar and Meeds within the contexts of a private league made up of high school and college friends, their findings may support our belief that fantasy sports participation is an inherently social activity. Whereas Lee at al. view motivations such as "game interest," "love for the sport," and "competition" as different from "social interaction," it may be that in contexts such as long-running leagues, these variables should be understood as inherently social. Similarly, the primary motivators identified by Farquhar and Meeds – arousal and surveillance – can be understood as emerging from the social nature of good-humored competition between friends and the desire to win the bragging rights.[32]

BALANCING WORK AND FANTASY "SNACKING"

The previous section demonstrated that fantasy sports participation can involve strong social elements and close friendships, especially when the same group of people plays in the same league over time. It is evidence that fantasy NBA is far more than simply an online game. This is also clear when respondents discussed the difficulties of playing fantasy sports amid everyday employment and work habits. For Australian NBA fans, this can be particularly difficult. During the NBA season the games begin between 10:00 AM and 2:30 PM Australian eastern standard time. If an enthusiast is not careful, score checking can drain the productivity out of any workday.

In the interview with husband and wife Al and Chanel, the topic of balancing fantasy sports research with work commitments was discussed in detail. Work-time media sports "snacking" was a vital part of effective fantasy NBA participation.[33] Snacking refers to short bursts of nonwork internet activity, usually involving social media websites, carried out at work. Australian fantasy NBA enthusiasts face various challenges in this regard. Not only does important research need to be undertaken during work time, but time-zone management becomes a measure of the successful manager. Australian enthusiasts balance their anticipation of the games beginning with work and family commitments (if I watch the box scores now, do I need to work late and therefore get home late?). Al described his experiences of balancing work commitments with fantasy NBA commitments:

> There are days where it would be a problem, but no, work is a higher priority. If I had something to do at work, that would obviously come first. . . . [I]f it's just the monotony of an everyday workday where I'm not on a really tight schedule, I will use looking at fantasy sports in a reward system–type setup. I'll have a browser window open with my team, and I might say, "I will write this next paragraph, then I'll look at the team". . . but if I think about the hours involved, and what I could have done with those hours . . . [34]

Research conducted by the consultancy firm Challenger, Gray & Christmas shows that fantasy sports can cost employers billions of dollars in lost productivity.[35] While Hakeem believed that there may be times when checking behaviors might enhance productivity, he also noted that shifting his focus back and forth between his work and sports scores was likely to limit his productivity:

> How can it not? The thing is it might be a good thing too because you can't just work flat out eight hours straight without some sort of a break. Maybe take a break every five minutes, but it's only for ten seconds. . . . I guess my mind's off work a little bit too much sometimes too. You're focused on that, and sometimes it is hard to refocus on your work. All that refocusing. If you didn't have to do it, you might be more productive. . . . How could [you] not?[36]

But other research suggests that snacking may even boost work-time productivity. Tussey argues that checking the scores while at work "may be as restorative and important to office cohesion as a trip to the water cooler."[37] Fantasy sports wife Linda, an academic in her early thirties, believed that her husband was most active in his fantasy sports league while at work:

> [It is] something that he does on the side to motivate him or keep him interested during the day. Even though his job is a really interesting job, it would break it up for him to do that. He works from home some days as well, so I'll hear him on those days talking sports with his friends. Generally, I will tune out because it's not something that I have a great interest in. Oh, he's talking stats again; he's talking sports again. You know, whatever.[38]

The desire to "tune out" when discussions of fantasy sports began was something that Chanel related to. Whereas Al believed that he thought about fantasy sports every "sixty-one seconds" or so (in the context that he recalled that men think about sex at least once every minute), his wife believed she often switched off – "That's my go- to re-sponse. . . . [Y]ou just strategically switch off, block it all out."[39] It is this issue of "tuning out" that guides our discussion of the role of gender in fantasy sports leagues.

FANTASY SPORTS LEAGUES AND THE PLAYING OUT OF GENDER ROLES

We found that "tuning out" was an experience shared by the fantasy sports wives in this study. Linda explained an adaptive behavior that she adopted, with some other wives and girlfriends of enthusiasts in this league, when her husband and his interleague friends were excitably discussing fantasy basketball: "We'd just tune out and talk about our own thing. Someone joked about starting a fantasy shopping website . . . where you collect a certain item like an expensive pair of shoes that people want and you can trade it to get the ultimate outfit or something [laughter], because that was the comparison for us. It [fantasy NBA] was a bit of a waste of time; it was silly. Why would you spend so much time doing that?"[40]

Carlos discussed the ways that his wife, Gabrielle, had worked to become more involved in his fantasy NBA interests. While he believed that it must be "testing" for her at times, he had long admired the ways that she had shown interest and been supportive of his enthusiastic participation in the league: "In the early stages I don't think she fully understood how deep we were in. . . . [A]s time progresses, and because she sees how much we invest in it, I think that her acceptance of it becomes greater. . . . It is not like these guys play it once in a while. This is what they look forward to. The meaning of draft day . . . It is really cool knowing . . . that your partner, my wife, gets it."[41]

Linda shared these beliefs. Although she would find her husband's involvement in fantasy sports "annoying and frustrating at times," she also believed that it was a sign of the maturity of her relationship that "you don't need to be involved in everything that they are doing. They can have separate interests that take up a lot of time." In this context, "tuning out" might be viewed as a strategy of support. Interestingly, Gabrielle – an executive in her late twenties who was portrayed as incredibly supportive by Carlos and described as having "a lot of exposure to this league" – also believed that tuning out was an important part of being a wife of a fantasy sports enthusiast: "I think I do just tune out. I try to stay involved for a little bit, to try and not be selfish, and try and be interested and engaged. But I think as soon as you start getting too deep into it is when I kind of know I can't contribute anything to the conversation [and] therefore [pause] *check out!*"[42]

Chanel has tried her best to understand Al's love of fantasy NBA, but she admitted to mostly experiencing confusion:

> Chanel: [When we were first dating] I didn't quite understand what you were trying to explain . . . He was trying to explain about chatting to your friends online, following the sports, and you have players, and I just kind of went, "I'm confused," so I would just smile and nod.
> Al: It was still the honeymoon phase, so I thought, "Break it now."[43]

Perhaps all male fantasy enthusiasts were stung by the suggestion of one of the most famous fantasy sports wives, Debbie, in the Judd Apatow film *Knocked Up.* After suspecting her husband, Pete, of being unfaithful, she follows him to a house in the suburbs where she discovers him participating in a live fantasy baseball draft. When she becomes angry, Pete defends himself – "It's a fantasy baseball draft. I'm not cheating or anything," to which Debbie replies, "No, this is worse."[44]

For Chanel, the meaning of fantasy sports was captured by the idea that *boys love to have the biggest numbers.* There certainly seems to be something quintessentially male about fetishistically watching and re-membering statistical representations of real sporting events. This was briefly addressed in Davis and Duncan's research and their respondent "Chris," who spoke with enormous pride about an impressive Allen Iver-son game for which he recounted the key statistics verbatim – "Iverson had 38 points, 12 assists, 5 steals and only 4 turnovers. . . . [H]e kicks so much ass for me!" This fits well with the first author's fond memories of Rajon Rondo having a massive triple-double on his wedding day. He was heard to have commented to one of his groomsmen, "How can this day get any better?" We believe that Hakeem described the feeling of statistical satisfaction particularly well: "It's all enjoyment. . . . I got into the statistical side of things, just by remembering everyone's stats. . . . [I]t was just that *feel of knowing.*"[45] The *feel of knowing* captures the embodied, and therefore gendered, dimensions of fantasy sports participation. It is not just a laborious, numerical, and time-consuming hobby. It makes you *feel* a certain way. Whoever has the biggest numbers has achieved something, if only personal satisfaction.

In the ESPN *30 for 30* documentary *Silly Little Game,* the inventors of rotisserie fantasy baseball were primarily interested in escapism and a way of being more involved in the sport they loved. Dan Okrent described feelings of boredom, loneliness, and emptiness that did not sit well with his "obsessive" personality. Davis and Duncan's research uncovered obsessive personality elements among enthusiasts. Davis confessed to being a roommate of a fantasy sports enthusiast and had witnessed him "repeatedly checking Web sites for statistics on several different players, bordering on the compulsive." The issue of obsession also emerged in the interview with Al and Chanel. Al believed that fan-tasy sports participation had "all the hallmarks of an addiction." Chanel, a medical psychologist, noted that fantasy sports reminded her of "the rat, pushing that little dopamine lever." The interviewer noted that in the lab, "that rat will push that bar until it starves to death." Linda also re-flected on the obsessive checking behaviors performed by her husband:

"When I think back on it, I will see him involved in looking up sports blogs as well – that's another thing that he checks. I don't really keep tabs because it is such a big part of his life. It is just something that he does *most* of the time really. There is always some sort of sporting check every hour or couple of hours."[46]

Many of the enthusiasts studied here expressed some version of the dilemma that is making life fit around fantasy NBA commitments. Indeed, the challenge for these enthusiasts is how they find the time to devote the effort required to manage an effective team, where a failure to "commit" to making the league competitive, and therefore fun, can result in unwanted social stigma. The interviewer posed such a scenario to Al and Chanel:

> Al: I make the analogy that we are like a military unit that
>   only survives because of how much we put in . . .
> Chanel: [mutters] Ridiculous.
> Al: . . . into the team effort. And what we do is only so good, becomes
>   so important to us, because we all invest so much. So the reaction
>   against those people who might switch off and not bother with
>   their team or so, it's just more disbelief. "How could they do that?
>   Look at all of this that we all get out of it. How could you?". . .
>   It's a bit of an affront to me because we're trying so hard.[47]

Laziness, weakness, lack of mental strength, and an unwillingness to stand up to one's wife or partner are all representative of this failure to commit, to let down the "unit." The ability to make decisions independent of wives and partners, to live one's life as one chooses, appeared as a cornerstone of manly independence in this league. As Hakeem demonstrated:

> Hakeem: I wanted to stay. But then my dad dragged me, because I didn't
>   want to go. My dad got me and said, "C'mon, you gotta go."
> Howie: Because you'd just been married in the previous weeks, or something.
>   Was your wife all right with it, though, or was she a bit upset?
> Hakeem: It was actually her birthday. But we had a birthday do the
>   next day, though, so, you know . . . fantasy day, I'm sorry! [Laughter]
>   It sounds a bit wrong, but she's the type of girl that isn't caught up
>   in all that shit. She won't crack the shits. . . . That's what I like about
>   her – she is *conscious* when it comes to importance in life.[48]

This is a fairly remarkable idea. For Hakeem, the love he felt for his wife was partly a result of her ability to understand the "importance" of fantasy sports events and how they were part of her husband's identity. Perhaps fantasy sports can be viewed as a proverbial *line in the sand* moment in the lives of fantasy sports enthusiasts and their partners. A test of one's manliness becomes increasing the time spent on fantasy sports and related activities without causing significant problems where one's wife begins to miss the time and attention that would otherwise be spent with her.

CONCLUSION: PROGRESS IN FANTASY SPORTS RESEARCH

Lee et al. argue that participation in fantasy sports provides a "mixed experience of participatory sport and spectator sport." They suggest that what one seeks through fantasy sports participation is the living of "a vicarious life."[49] But we argue that there is nothing vicarious about it. On the contrary, fantasy sports are an incredibly important part of the real, everyday lives of the dedicated fantasy enthusiasts studied here. The importance of fantasy sports in the lives of these enthusiasts was seemingly understood by each of their wives. As such, fantasy sports became important in their lives, too. Fantasy sports occupy gendered terrains, where hegemonically masculine sporting conditions are, at least partially, reproduced. But they are reproduced in complex ways involving social groups of women and men, sports media technologies, sporting practices and masculine behaviors, workplace habits and rituals, and loving relationships born of heteronormative social privilege.[50]

The respondents were encouraged by the authors to talk about what fantasy sports meant to them. Based on this conversational style, three broad and interrelated issues emerged. First, fantasy sports, when played over long periods in one league with the same members, are an intrinsically social experience. We are mindful of other research that challenges the idea that fantasy sports are mostly about socializing. We argue, however, that many of the factors identified as nonsocial by Lee et al. and Farquhar and Meeds may, after further investigation, be revealed as social in nature. The differences in our conclusions may be related to certain methodological differences between our study and theirs. In Farquhar and Meeds's, respondents were recruited on a college campus, a decision that resulted in a younger group of participants. The participants in our study were once college students too, so the differences in our findings may be explained by factors such as length of time playing fantasy sports and demographic elements such as income and family status. It may be that as respondents grow and mature, their understanding of why they play fantasy sports may change. Lee et al. recruited from online message boards.[51] It may be that when one's league fails to provide adequate social stimulation, fantasy sports enthusiasts will seek out alternative forms of social engagement. Many in their respondent group were possibly seeking to engage in the social aspects of fantasy sports through these message boards at the time they were recruited.

Second, fantasy sports participation must be balanced with ordinary working habits. This aspect of our findings is undoubtedly linked to the time-zone issues that Australian enthusiasts of NBA basketball face. Games are played between 10:00 AM and 2:30 PM on workdays. Watching scores, picking up hot free agents, and making trades are activities that occur during work time for those in traditional 9:00–5:00 industries and sectors. Given that demographic evidence suggests that many enthusiasts are professionals, this may be a common dilemma in some geographical regions. Is it a refreshing respite, like standing near a water cooler? Is it a productivity drain? This may be a field of further study that will have significant implications in not only sports media research but also business and human resource management fields.

Third, women are active participants in fantasy sports leagues. While most research suggests that there is little role for women in fantasy sports, citing evidence that only around 4 percent of league participants are women, we argue that this oversimplifies the roles that women play. The men who play fantasy sports in our study are supported by wives and partners. Without the understanding of women, the fantasy sports industry may be far less influential and widespread. The context for this claim is that fantasy sports continue to be played long after the computer is switched to sleep mode. Men discuss fantasy sports at social events, while watching sports, over the phone, via e-mail, and with their wives and partners. The women in this study reported "tuning out," but this should not be misunderstood. Tuning out appears as a way of understanding and appreciating something significant in their loved ones' lives. The women in this study don't fully understand their husbands' devotion to such a silly little game, but they have no desire to change it.

There is little empirical research exploring the social meanings of fantasy sports participation, and there has not been, before this chapter, empirical research examining the roles and perspectives of fantasy sports wives and partners. It is a paucity that we have sought to address in these pages. In this emerging field of study, the imperative is for researchers to work together across methodological divides. We should collaborate to cultivate the kinds of eyes that will unlock the social potential of fantasy sports leagues. It may be that future collaborations and further research will use mixed methods, drawing on the collective skill sets of fantasy sports researchers. We also believe that researchers should learn to play, love, and live fantasy sports. It will provide an important perspective on what is emerging as the data continue to flow.

NOTES

1. Fantasy Sports Trade Association, "Fantasy Sports Participation Sets All-Time Record, Grows Past 32 Million Players," June 10, 2011, http://www.fsta.org /blog/fsta-press-release/fantasy-sports -participation-sets-all-time-record-grows -past-32-million-players; James Montague, "The Rise and Rise of Fantasy Sports," CNN, January 20, 2010, http://www.cnn .com/2010/SPORT/football/01/06 /fantasy.football.moneyball.sabermetrics /index.html.

2. Seunghwan Lee, Won Jae Seo, and B. Christine Green, "Understanding Why People Play Fantasy Sport: Development of the Fantasy Sport Motivation Inventory (FanSMI)," *European Sport Management Quarterly* 13 (2013): 166–199.

3. Luke Esser, "The Birth of Fantasy Football," *Fantasy Index*, 1994, https:// www.fantasyindex.com/resources/the -birth-of-fantasy-football; Frank M.

Shipman, "Blending the Real and Virtual: Activity and Spectatorship in Fantasy Sports," 2001, http://www.csdl.tamu .edu/~shipman/papers/daco1.pdf; Jim Hu, "Sites See Big Season for Fantasy Sports," *CNet*, August 8, 2003, http://news.cnet .com/2100-1026_3-5061351.html.

4. Amber Smith, David P. Synowka, and Alan D. Smith, "Exploring Fantasy Sports and Its Fan Base from a CRM Perspective," *International Journal of Business Innovation and Research* 4 (2010): 103–142; Nicole Davidson, "Internet Gambling: Should Fantasy Sports Leagues Be Prohibited?," *San Diego Law Review* 39 (2002): 201–268; Richard T. Karcher, "The Use of Players' Identities in Fantasy Sports Leagues: Developing Workable Standards for Right of Publicity Claims," *Penn State Law Review* 111 (2007): 557–585.

5. Hunter S. Thompson, *Fear and Loathing in Las Vegas: A Savage Journey to*

the Heart of the American Dream (London: Harper Perennial, 2005), 68; Hunter S. Thompson, *The Gonzo Papers Anthology* (London: Picador, 2009), 1062; Catherine Kohler Riessman, *Narrative Analysis* (Newbury Park, CA: Sage, 1993); Donna Haraway, *Modest_Witness@Second _Millennium.FemaleMan©_Meets _OncoMouse™* (New York: Routledge, 1997); Donna Haraway, *How Like a Leaf: An Interview with Thyrza Nichols Goodeve* (New York: Routledge, 2000), 107.

6. Loic Wacquant, *Body and Soul: Notebooks of an Apprentice Boxer* (New York: Oxford University Press, 2004), xi.

7. Terry Flew, *New Media: An Introduction* (South Melbourne: Oxford University Press, 2005), 5–7.

8. World Fantasy Games, "Fantasy Sports Demographics," 2011, http://www .worldfantasygames.com/site_flash /index-3.asp.

9. Nikolas W. Davis and Margaret Carlisle Duncan, "Sports Knowledge Is Power: Reinforcing Masculine Privilege through Fantasy Sport League Participation," *Journal of Sport and Social Issues* 30 (2006): 244–264.

10. Sherry Turkle, *Life on the Screen: Identity in the Age of the Internet* (London: Weidenfeld & Nicolson, 1995); Robert M. Kitchin, "Towards Geographies of Cyberspace," *Progress in Human Geography* 22 (1998): 385; Donna Haraway, *Simians, Cyborgs, and Women* (London: Free Association Press, 1991).

11. Michael A. Messner, Margaret Carlisle Duncan, and Kerry Jensen, "Separating the Men from the Girls: The Gendered Language of Televised Sports," *Gender and Society* 7 (1993): 121; Toby Miller, *Sportsex* (Philadelphia: Temple University Press, 2001).

12. R. W. Connell, *Masculinities* (Crows Nest, NSW: Allen & Unwin, 2005), 54,

59; Bryan S. Turner, *The Body and Society: Explorations of Social Theory* (Oxford: Blackwell, 1984).

13. Toni Bruce, "Reflections on Communication and Sport: On Women and Femininities," *Communication and Sport* 1 (2013): 128.

14. Davis and Duncan, "Sports Knowledge Is Power," 245; Luke Howie and Perri Campbell, "Privileged Men and Masculinities: Gender and Fantasy Sports Leagues," in *Digital Media Sport: Technology, Power, and Culture in the Network Society,* edited by B. Hutchins and D. Rowe (New York: Routledge, 2013), 235–248.

15. Davis and Duncan, "Sports Knowledge Is Power," 251; Howie and Campbell, "Privileged Men and Masculinities," 242–245.

16. Davis and Duncan, "Sports Knowledge Is Power," 251.

17. Although women's participation as team managers may be trending upward, see Brett Hutchins and David Rowe, *Sport beyond Television: The Internet, Digital Media, and the Rise of Networked Media Sport* (New York: Routledge, 2012), 173–175; and Jameson Otto, Sara Metz, and Nathan Ensmenger, "Sports Fans and Their Information-Gathering Habits: How Media Technologies Have Brought Fans Closer to Their Teams over Time," in *Everyday Information: The Evolution of Information Seeking in America,* edited by William Aspray and Barbara M. Hayes (Cambridge: MIT Press, 2011), 185–216.

18. Howie and Campbell, "Privileged Men and Masculinities," 246; Jim McKay, Michael A. Messner, and Don Sabo, "Studying Sport, Men, and Masculinities from Feminist Standpoints," in *Masculinities, Gender Relations, and Sport,* edited by Jim McKay, Michael A. Messner, and Don Sabo (Thousand Oaks, CA: Sage, 2000), 1–12; Michael A. Messner and Don Sabo,

eds., *Sport, Men, and the Gender Order: Critical Feminist Perspectives* (Champaign, IL: Human Kinetics, 1990).

19. Kathleen Ervin, "Women against Fantasy Sports," *Failure Magazine*, 2012, http://failuremag.com/index.php/feature/article/women_against_fantasy_sports/.

20. Davis and Duncan, "Sports Knowledge Is Power," 247.

21. Lee, Seo, and Green, "Understanding Why People Play Fantasy Sport," 174.

22. Davis and Duncan, "Sports Knowledge Is Power"; Howie and Campbell, "Privileged Men and Masculinities," 237–238.

23. Lee K. Farquhar and Robert Meeds, "Types of Fantasy Sports Users and Their Motivations," *Journal of Computer-Mediated Communication* 12 (2007): 1212, 1208.

24. Ibid., 1212; Lee, Seo, and Green, "Understanding Why People Play Fantasy Sport," 171, 170.

25. Interview with Al (enthusiast), July 24, 2013.

26. In truth, it was slightly more complex than this.

27. Davis and Duncan, "Sports Knowledge Is Power," 256; interview with Al (enthusiast), July 24, 2013.

28. Interview with Al (enthusiast) and Chanel (fantasy sports wife), July 24, 2013.

29. Interview with Hakeem (enthusiast), July 27, 2013.

30. Lee, Seo, and Green, "Understanding Why People Play Fantasy Sport"; Farquhar and Meeds, "Types of Fantasy Sports Users and Their Motivations."

31. Interview with Carlos (enthusiast), July 30, 2013.

32. Lee, Seo, and Green, "Understanding Why People Play Fantasy Sport," 166; Farquhar and Meeds, "Types of Fantasy Sports Users and Their Motivations," 1208.

33. E. Tussey, "Desktop Day Games: Workspace Media, Multitasking, and the Digital Baseball Fan," in *Digital Media Sport: Technology, Power, and Culture in the Network Society,* edited by B. Hutchins and D. Rowe (New York: Routledge, 2013), 43.

34. Interview with Al (enthusiast), July 24, 2013.

35. Drew Guarini, "Fantasy Football Costs Upwards of $6.5 Billion, Study Finds," *Huffington Post,* September 4, 2012, http://www.huffingtonpost.com/2012/09/04/fantasy-football-costs-employers_n_1855492.html.

36. Interview with Hakeem (enthusiast), July 27, 2013.

37. Tussey, "Desktop Day Games," 43.

38. Interview with Linda (fantasy sports wife), July 22, 2013.

39. Interview with Chanel (fantasy sports wife), July 24, 2013.

40. Interview with Linda (fantasy sports wife), July 22, 2013.

41. Interview with Carlos (enthusiast), July 30, 2013.

42. Interview with Linda (fantasy sports wife), July 22, 2013; interview with Gabrielle (fantasy sports wife), July 30, 2013.

43. Interview with Al (enthusiast) and Chanel (fantasy sports wife and enthusiast), July 24, 2013.

44. Judd Apatow, *Knocked Up* (Los Angeles: Apatow Productions, 2007).

45. Davis and Duncan, "Sports Knowledge Is Power," 253; interview with Hakeem (enthusiast), July 27, 2013.

46. Adam Kurland and Lucas Jansen, *Silly Little Game,* season 1, episode 11, ESPN *30 for 30* documentary series (ESPN and Team Marketing, 2010); Davis and Duncan, "Sports Knowledge Is Power," 253; interview with Al (enthusiast) and Chanel (fantasy sports wife and enthusiast), July 24, 2013; interview with Linda (fantasy sports wife), July 22, 2013.

47. Interview with Al (enthusiast) and Chanel (fantasy sports wife and enthusiast), July 24, 2013.

48. Interview with Hakeem (enthusiast), July 27, 2013.

49. Lee, Seo, and Green, "Understanding Why People Play Fantasy Sport," 166, 170.

50. See also Howie and Campbell, "Privileged Men and Masculinities"; and Davis and Duncan, "Sports Knowledge Is Power."

51. Farquhar and Meeds, "Types of Fantasy Sports Users and Their Motivations"; Lee, Seo, and Green, "Understanding Why People Play Fantasy Sport."

FIVE

Domesticating Sports: The Wii, the Mii, and Nintendo's Postfeminist Subject

Renee M. Powers and Robert Alan Brookey

IN 2005 NINTENDO BEGAN RELEASING INFORMATION ABOUT their next console, code-named "Revolution." The reception from the video game press was rather mixed. Ryan Block, covering Nintendo's introduction of the Revolution at the 2005 Electronic Entertainment Expo (E3) for the tech blog *Engadget,* had this to say: "The Revolution is a really unsexy device, all things considered – but it is a prototype, and [Nintendo] did hammer home that they want input from their adoring public. This may also just prove that Nintendo is serious when they say they don't care about the hardware as much as they do about the gaming experience. They had to show something, and they did. It didn't hurt them, it didn't help them." Mark Casamassina, writing for *IGN,* provided a more positive assessment: "At E3 2005, Nintendo unveiled the Revolution console. It is the company's sleekest unit to date. The tiny-sized system is designed to be quiet and affordable. The revolutionary aspect of the machine – its input device – remains a secret."[1] Yet even Casamassina noted how the new console broke with industry tradition by not incorporating significant technological advances in graphic capability.

This break from the industry, however, was actually a point of pride for Nintendo. The company took every opportunity to announce that its approach to their new products was part of a "Blue Ocean Strategy" (BOS). Perrin Kaplan, vice president of marketing and corporate affairs for Nintendo of America, offered this explanation: "Seeing a Blue Ocean is the notion of creating a market where there initially was none – going out where nobody has yet gone. Red Ocean is what our competitors do – heated competition where sales are finite and the product is fairly predictable. We're making games that are expanding our base of consumers in Japan and America."[2] The BOS was a business model developed by W. Chan Kim and Renée Mauborgne, and they used the term *blue oceans* to "denote all industries not in existence today – the unknown market space, untainted by competition." Later they would expand on this strategy to identify its specific features: "Create uncontested market space. Make the competition irrelevant. Create and capture new demand. Break the value/cost trade off. Align the whole system of a company's activities in pursuit of differentiation and low cost."[3]

Indeed, in order to break the value-cost trade-off, Nintendo was taking a very different direction from its competitors. While Nintendo was promoting the Revolution, Microsoft was touting its next console, the Xbox 360, and Sony was promoting the forthcoming PlayStation 3. Both the PS3 and 360 were designed to offer significant technological advances that would operate on HDTVs, whereas the Revolution would include only a standard DVD drive.[4] Yet while the other companies were focusing on advancements in video technology, Nintendo worked on advancing the experience of video game play. When Nintendo finally released its new console, under the Wii brand name, it highlighted the new wireless remote control pod (the "secret" that Casamassina alluded to) that could track movement in three-dimensional space. In addition, the Wii sold for $249, whereas the Xbox 360 and the P3 retailed for significantly more. Because it did not incorporate the expensive technology of the other competing consoles, Nintendo's Wii offered an enhanced gaming experience, without significantly increasing the price of its product.

Due in part to its price point and its unique Wii Mote (a wireless controller with motion-sensor capability), the Wii console caught on with consumers. During the holiday shopping season of 2006, it was the "must-have" present, and stores would empty their stock as soon as new shipments of the console arrived.[5] In addition, the Wii console came with the *Wii Sports* game package preloaded. This game package fully utilizes the tracking functionality of the Wii Mote, in such sports simulation games as bowling, baseball, tennis, golf, and boxing. While the Wii console was topping the charts in hardware sales, the *Wii Sports* package was topping the charts of video game software sales. Furthermore, the *Wii Sports* package was a precursor to another important product, the

Wii Fit, which was released about a year after the Wii. The *Wii Fit* was a peripheral add-on to the Wii console, with a corresponding software package. It includes a balance board that communicates with the Wii console and also tracks the movement of the player. The corresponding software package includes some sports simulation games, such as snowboarding and skiing, but also includes fitness exercises like yoga.

Another important characteristic of the Wii is the Mii avatar. The Wii console allows players to create characters, very cartoonish in appearance with large heads and small bodies, and these function as avatars in both the *Wii Sports* and the *Wii Fit* programs. In fact, in the *Wii Fit* program, not only does the player use the Mii to interact with the different games as fitness exercises, but the Mii may also be used to track the player's progress in relation to weight loss goals. The ultimate function of the Mii, however, is to link the Wii console, the *Wii Sports* package, and the *Wii Fit* package by allowing the player to use the same avatar across these products.

Since their releases, the Wii console, *Wii Sports,* and *Wii Fit* have racked up some impressive numbers. According to Nintendo's own sales figures, the Wii has sold ninety-seven million units, while *Wii Sports* (including the *Wii Sports Resort* package) has sold more than one hundred million units, and *Wii Fit* (and *Wii Fit Plus*) has sold forty-five million units. Although demand for the Wii has waned over the years, it still commands 41 percent of the global video game market, a significant improvement over its predecessor, the GameCube.[6]

Nintendo's marketing strategy has proved successful, particularly in moving beyond the established target market for video games, and a good deal of that success was realized by cultivating women as new consumers. Indeed, although others have tried to market video games to women, their success has been very limited, and Nintendo has succeeded where others have failed.[7] Not only was the Wii designed with women in mind, but women were also featured in Nintendo's marketing campaign. It would be easy to consider the Nintendo Wii as an important moment for women and the gaming industry, because it successfully welcomed them into a market dominated by men, and it did so with a sports simulation program, again a market often dominated by men. But we maintain that Nintendo's success does not necessarily bode well for women.

Nintendo's efforts to design and market the Wii and *Wii Fit* to women reveal a postfeminist view of gender, one that reconstitutes body image and appearance as of primary importance to women. In addition, Nintendo imagines women to be mostly domestic creatures whose identity is formed around their roles as wives and mothers. While the Wii may seem to invite women to participate in sports activities, it domesticates those sports and anchors them to stereotypical concerns about feminine body image. We begin with an explanation and definition of postfeminism, which will focus on the specific tenets that are applicable to our analysis. We then will look at specific details of Nintendo's efforts to design and market the Wii, the *Wit Fit,* and the *Wii Sports* packages to show how they also reflect postfeminism.

DEFINING POSTFEMINISM

Our definition of postfeminism, as informed by Diane Negra, relies on three tenets.[8] First, postfeminism is the theory that regards all of feminism's goals regarding gender equality as achieved. Second, postfeminism relies on a consumer-centric, appearance-obsessed society to thrive. It is in this way that female empowerment is oriented toward consumer choice, favoring free-market capitalism. Third, underlying economic choice is the conservative rhetoric of individualism, which also favors traditional family values. The correct choices under a postfeminist framework rely on individualistic traditional femininity. For instance, postfeminism understands "home" not in a utilitarian sense, but rather as a space to display a woman's aesthetic and consumer choices, as dictated by commercialized media. The "choices" of domestication and marriage are considered two of the most valued characteristics in postfeminism. The social and consumer choices proffered to women through postfeminism are framed to emphasize women's traditional roles as attractive wives and dutiful mothers.

The call of postfeminism is not simply to consume, but to consume correctly. For example, Negra makes the point that aging is distasteful in postfeminism, as patriarchal ideals of femininity dictate women must always look young and sexually attractive to men. But women can combat the process of aging through a variety of consumer choices. Women have the choice to use various antiwrinkle creams, undergo cosmetic surgeries such as breast augmentation, and subject themselves to Botox or collagen injections in the attempt to hide or reduce the effects of aging. Although it is true that women can choose whether to combat the body's changes as it ages, postfeminism is preoccupied with presenting the right choices as those that combat age. Postfeminism contends that all women are similar or "bound together by a common set of innate desires, fears, and concerns."[9] One of these innate desires is the importance of the aesthetic of youth or at least looking younger than one's age, which can be achieved through the correct purchases.

Negra also combines domesticity and body grooming as similar markers in postfeminist notions of authentic or "innate" femininity. This authentically feminine self can be achieved through purchasing the correct products as well.[10] The authentic self is "a *sexier* self, in which sexual attractiveness has been magically transformed from an oppressive imperative of patriarchy into a source of power *over it,* a brave new postfeminist self requiring continual self-monitoring and investment in salons and spas, fashion stores, and regular visits to the gym."[11] Negra suggests that such self-monitoring, while working to achieve authentic femininity, is one of the most distinctive features of postfeminism.[12] This includes striving for the culturally determined ideal female body as well as keeping the perfect home while making the work look effortless. The essential postfeminist female nurtures her family through her efforts within the home: cooking gourmet meals, decorating an enchanting living room, and keeping her bikini line waxed. Indeed, these ideals are often determined in accordance with male heterosexual desire. This ideology relies on strictly fixed gender roles. Thus, postfeminism renders female choice as empowering consumption, only in accordance with patriarchal standards.

Yvonne Tasker and Diane Negra identify the connection between consumerism and individualism. The turn of the twenty-first century saw a repackaging of feminism as simply a part of Western culture; young women have grown up with feminism in the water. Acknowledging the existence of feminism leads to "a prepackaged and highly commodifiable entity so that discourses having to do with women's economic, geographic, professional, and perhaps most particularly sexual freedom are effectively harnessed to individualism and consumerism."[13]

The problem with a society permeated by postfeminism is the way in which popular culture persuades women to make these traditional choices with the understanding that individual choices are empowering. Negra typifies popular women-centered cinema as stories of "miswanting" or "narratives of adjusted ambition."[14] Negra claims chick flicks perpetuate the idea that a career is not necessarily what the female character wanted in the first place. That is, in fact, what she "miswanted." Instead, the female protagonist realizes by the end of the film that she wants marriage and family. Often, though, a career will lead the protagonist to romance; thus, her career is deemed necessary but only as a means to an end. Once a romantic partner is secured, the woman's true calling can be claimed: domestication. In this example, a woman's individual choice to pursue a career is renegotiated to reinforce traditional femininity: settling down, partnering, and raising a family. This narrative is delivered as though it is an authentic and empowering decision.

Correspondingly, the female worker in cinema is celebrated, but often the film "reconfigures the ideological underpinnings . . . in many cases, to reaffirm the centrality of heterosexual marriage."[15] Whereas feminist notions of work allowed women to free themselves from the necessary institution of marriage, postfeminist notions of work insist that women can now marry for love instead of money. Postfeminism persuades women *to* marry, as marriage is the ultimate goal. Furthermore, the discussion of female workers no longer is a feminist discussion about finding good work, but rather a postfeminist discussion on how finding any work may hinder marital happiness. Again, the choice to stay home and raise a family is rendered empowering and can be another example of a woman's "miswanting."

The three tenets that we have outlined do not act independently of one another, nor are they the only forms of postfeminist rhetoric. Commercial consumption, sexual allure, and domesticity, however, are central to postfeminism and are important to our analysis. Indeed, while the Wii console, the *Wii Sports* package, and the *Wii Fit* have been instrumental in opening the male-dominated market of video gaming to women, they do so in ways that focus on women's domesticity and body image.

Although Nintendo is credited with opening the video game market to women, it is worth noting the company's responsibility in creating the male-dominated market that video games had become. Starting with the NES in 1985, Nintendo carefully targeted their products to a specific core demographic of males between the ages of eight and fourteen. In fact, they had three main strategies that were designed to target this demographic. First, the company conducted extensive, ongoing premarket testing of its games with this demographic. Second, they published the magazine *Nintendo Power*, which was devoted to features about new games for Nintendo's systems and advertisements for those games. Finally, Nintendo used its call centers to create an information feedback loop for game development. When players called Nintendo to either register a complaint or comment on the game or seek help, those calls were collated as data for future game design.[16]

When other companies entered the video gaming market, they would often target a demographic that was older than Nintendo's target market. For example, Sega targeted males between the ages of fifteen and seventeen, and Sony targeted post–baby boomers who were familiar with the Sony brand, having grown up with products like the Walkman. When Microsoft entered the market with the Xbox, it targeted males between eighteen and thirty-four.[17] Nintendo's response to this competition was to double-down on its core market of young males, but they soon found themselves outflanked by the competition. When the GameCube was introduced, Nintendo found that its core demographic was reduced to a much smaller share of the video game market.

While Nintendo has been aggressive in getting its products extensive media coverage, it has been very restrictive in allowing the media access to its operations. Perhaps due in part to the adoption of their BOS, Nintendo eased those restrictions in order to let *Nikkei Business,* a prominent Japanese business magazine, publish a series of articles about the practices behind the design of the Wii. These articles were then published as a book by *Nikkei Business* reporter Osamu Inoue, entitled *Nintendo Magic: Winning the Videogame Wars.* We should note that the book is a very flattering chronicle of Nintendo's development practices and is equally adulating of their executive team – so much so that it is easy to conjecture that a favorable portrait was one of Nintendo's conditions for access. Nevertheless, Inoue gives a quite detailed account of how the Wii was designed and perhaps the only report that is based on direct access to Nintendo's management.

As Inoue notes, Nintendo broke from the rest of the video game industry in the drive to up technological performance, and instead "the new system would expressly seek out technology that would endear itself to families – a 'Mom has to like it' approach to development." He goes on to note that "a basic tenet of the Wii's design was mom-friendliness." It was Nintendo executive and legendary game designer Shigeru Miyamoto, however, who defined the influence of women's attitudes in more specific terms: "the Wife-o-meter." In other words, Miyamoto determined the success of the console's design, and its software, by how well his wife responded to the products. As Inoue explains, "Miyamoto's wife had never had any interest in videogames. . . . When the Wii came out, she turned it on of her own accord and used the Mii character creator to make caricatures of her family and friends."[18]

Indeed, the interest of wives and mothers can be seen at every level of design of the Wii, from the hardware to the software and to the peripherals. For example, the Wii console was purposely designed to be small and take up less space and be less visually obtrusive in the living room, therefore not marring the domestic aesthetic. Because the console did not have the same processor capacity of its competitors, it would also use less electricity and generate less heat, thereby forgoing the need for noisy cooling fans. As Nintendo president Satoru Iwata argued, "You can't have the fan run while it's in standby at night. If Mom hears the fan, she'll pull the plug." In addition, the Wii was specifically designed to be backward compatible. As Inoue explains, "Up until the Wii, none of Nintendo's home consoles had been backward compatible; if you wanted to play an old game, you had to keep the old console, and Mom couldn't stand seeing all those videogame systems lined up by the TV."[19] With the sleek and quiet Wii, mothers and wives had no reason to complain.

However, Nintendo's targeting of women went beyond simple aesthetic design.

The Wii Mote was also designed with women in mind, and Iwata is credited with choosing a wireless controller because it did not have the aesthetically unpleasant wires. Furthermore, the decision to call the controller a Wii Mote was a purposeful decision to relate it to the television remote, and therefore make it seem less intimidating to members of the family who were not video game players. In addition, the interface of the Wii operating system turned game programs into "channels," and the design offers these channels to players as if they were television channels. Inoue explains, "The TV was common property in a typical household. There were programs that everyone watched together. . . . The goal for the Wii was the same – for it to have something for everybody in the family, for it to be the channel that someone would want to watch every day."[20] The Wii interface has "channels" for news, weather, photos, and shopping, and when new games are loaded on the system, they too appear as channels on the interface. In addition, these channels appear on the interface as small television screens within the larger screen of the actual television to which the console is connected.

Although the design rationale for the Wii seemed to focus on the family, the emphasis was more finely tuned. There was a very specific gender assumption that imagined that the wife and mother is central to the family and that pleasing her was of primary importance. Therefore, Nintendo mainly reached out to the women who occupy very traditional gender roles, and these marketing efforts are also quite apparent in the advertisements for the Wii. Although the early Wii advertisements featured a pair of Japanese men traveling the countryside telling various people "Wii would like to play," women were prominent in the ads and were usually portrayed in family contexts. For example, when the Wii men visit a house in a suburban setting, a father with his two children answer the door, while the mother peeks over their shoulders in the background. But soon she assumes the foreground and seems to dominate the game as she bowls and then jogs in a running game. At another house, the Wii men encounter a different family, but the scene plays out in a similar manner. This time a young man in the family (possibly the son) loses a boxing match with an older woman (possibly the mother), who wins and jumps about victoriously. Not only do these scenes depict these women playing the video games, but the second scene also shows the older woman beating the younger man at a video game version of the rather decidedly masculine sport of boxing. This focus on females is carried over to the shorter thirty-second TV spot, which depicts only the first scene that puts the mother in the center of the action.

We should note that not all women who appear in the Nintendo Wii ads are depicted as wives and mothers, but it is important to point out how these ads reflect the design strategy of the console. The commercials demonstrate that while the Wii is a console for the whole family, it is a product that wives and mothers will approve of and enjoy. Furthermore, these commercials also focused on the games in the *Wii Sports* package, and the women are depicted as competent at these sports games, even when it involves masculine pursuits such as boxing. These depictions might be regarded as feminist, to the degree that they show women competing and winning at traditionally masculine sports, but the focus on these women as wives and mothers is a decidedly postfeminist constraint.

The *Wii Sports* package included several games for which women already enjoy recognition, such as golf and tennis. What is interesting about the *Wii Sports* package, however, is that although it is a sports simulation game, in keeping with Nintendo's BOS, it departs significantly from most sports simulation games. Most competing games attempt to offer the player an authentic experience, as close to the actual sport as possible. Indeed, many sports simulation games are branded by the professional organizations that oversee these sports, the NBA, NCAA, NFL, FIFA, and so on. These games also incorporate actual teams and actual players into the simulated gameplay, and sometimes actual stadiums are reproduced as the scenes for the games. *Wii Sports,* on the other hand, does none of these things. The sports take place in vibrantly colorful, cartoonish settings that are abstracted from any real-world sports venues. The game players are the cute, cartoonish Miis created by the game players, and the only brand to be found is Nintendo.

Whereas other sports simulation games often call on an acumen regarding the specific sports cultures, the players and the teams (if not actual physical powers), the *Wii Sports* package does not. In fact, the games require little knowledge beyond a rudimentary understanding of the sports. Although the games require physical movement, they do not necessarily require physical rigor. If anything, the Wii levels the playing field where physical strength and dexterity are concerned. In fact, one of the important uses of the Wii has been in physical therapy treatments and in nursing homes, because the games are easily played by individuals whose mobility has been impaired because of an accident, surgery, or advancing age.[21] The Wii lets everyone play and makes it possible for everyone to win, and if the commercial is to be believed, it even allows a mother to beat her son at boxing. Whereas other sports simulations heighten the competition and public trappings of professional sports, *Wii Sports* domesticates its activities into family-friendly games that even Mom can play. The program does not celebrate women as sports figures, but rather reflects an assumption that sports need to be feminized for women, abstracted from any knowledge of actual sports culture, with the physical demands reduced in order for women to be competitive.

"THAT'S OBESE"

About a year after the release of the Wii, the *Wii Fit* application was introduced, which included a balance board and several health and exercise software programs. Although the ad campaign for the *Wii Fit* represented both genders, women figured more prominently in the campaign. For example, a *Wii Fit* television spot includes four women and only three men. This would not seem to be an important disparity, but this focus on women would carry over in both the packaging and the print promotion for the *Wii Fit*. The box art for the Japanese release of the application would feature the silhouette of a woman holding a yoga position. An original ad image that featured both men and women was reconfigured for a Best Buy circular so that one of the women dwarfs the men. Finally a promotional poster shows only a woman, with the tagline "How will it move you?" It is important to note that the *Wii Fit* was an add-on to the original console, so anyone who purchased this add-on had to either own the console or purchase it as well.

In fact, when a player starts up the *Wii Fit*, she must import an existing Mii to the *Wii Fit* Plaza, the game's navigation screen. An animated balance-board character introduces itself and explains, "I'll be here to help you achieve your fitness goals." The user is prompted to enter his or her height and date of birth, which will be used to calculate body mass index (BMI) and *Wii Fit* Age. After an initial fitness test, the balance-board character introduces the user to the types of training activities available: Yoga, Strength, and Aerobics.

The yoga and strength-training activities are led by a male or female trainer, as selected by the user. In these activities the trainer takes the player through several poses (half-moon, warrior, sun salutation) and movements (torso twist, lunge, jackknife). The aerobic activities do not feature a trainer, but they also most closely resemble the other simulation games found on the *Wii Sports* package. Only one, boxing, is carried over from the package, but the others, including the running, skiing, snowboarding, and soccer activities, reproduce the brightly colored environments found in *Wii Sports,* and the Miis, of course, are the same.

A returning *Wii Fit* user will be greeted by the animated balance board, which will identify how long it has been since the user's last workout. It then asks if the user would like to hear a fitness tip. These tips range from general health tips, such as the importance of regular exercise, to game-specific tips, like how to change the trainer within certain activities. Each time a Mii is selected, the user must engage in a general fitness test, the results of which determine the user's *Wii Fit* Age. This general fitness test requires the user to identify how heavy his or her clothes are (the options are two pounds, four pounds, or custom), then the balance board weighs the individual, determines BMI, and performs a center-of-balance test. The balance-board character explains that the *Wii Fit* Age is determined by averaging these scores with the user's reported age.

After a user's *Wii Fit* Age is determined, the user is prompted to set weight loss goals. The user indicates how many pounds he or she would like to lose and a deadline to complete the goal. If the goal is not reached by the deadline, the balance-board character says, "Maybe next time you should shoot for a smaller goal, one that's easier to reach?" regardless of how small the goal was.

The *Wii Fit* reports a user's BMI on a chart that indicates whether the user is underweight, normal, overweight, or obese. If the user's BMI is normal, the player's Mii cheers, and the balance-board character suggests the user try to reach the ideal BMI of 22. If the user's BMI is underweight, the player's Mii cheers and the balance-board character suggests using the *Wii Fit* training to gain muscle mass. Interestingly, if the user's BMI is determined overweight or obese, the user's Mii hangs its head in shame as its waist enlarges. Adrienne Massanari indicates that this is one of the ways the *Wii Fit* disciplines players' bodies, as the player's self-representation is no longer within the user's control.[22] The balance-board character explains that a BMI over 25 puts the user at a higher risk for health issues. As the user finishes his or her fitness test, the balance-board character ends the session with, "Your BMI indicates you are obese, so focus on watching what you eat and lowering your BMI, [username]."

When a user's fitness tests indicate an overweight or obese BMI, the user's Mii swells, reflecting a significant weight gain. However, this is the only characteristic that becomes customized. Though the *Wii Fit* asks for the user's height, the Mii of a user who indicates his or her height as six foot one is represented as the same height as the Mii of a user who indicates his or her height is four foot ten. However, a Mii with a BMI of 25 is significantly thinner than a Mii with a BMI of 35. The differences in weights and similarities in heights are startling when the Miis stand next to one another in the *Wii Fit* Plaza. This emphasizes the *Wii Fit*'s commitment to ideal standards of weight. Yet perhaps the most important difference is the way the *Wii Fit* treats individuals with different BMIs: the underweight are cheered, and the overweight are a source of shame.[23]

Despite repeated references to fitness and posture, the dominant theme of the *Wii Fit* is weight loss. When a new user is introduced to the platform, the balance-board character explains the importance of good posture: poor posture indicates a lack of balance due to a sedentary lifestyle, leading to fat accumulation and the potential for health problems. This is only one of many references to posture as a euphemism for weight loss. The balance board shows the user an illustration of good posture versus poor posture. Both bodies shown are very thin women, presumably white, wearing sports bras and cropped fitness pants, and these bodies are similar to the ones featured in the ads mentioned earlier. The illustration of poor posture also includes an oversize handbag over her right shoulder. The balance-board character continues by explaining the importance of good core muscles in relation to good posture, which "can raise your base metabolism and help you burn fat more efficiently." Here we recognize the direct connection between posture and weight loss. The illustration of this thin woman can also serve as an example of "thinspiration," or images of thin women used as inspiration or motivation to maintain a diet or exercise plan (or both).

When the balance-board character introduces a new user to the training activities, it mentions the benefit the *Wii Fit* has on flexibility (yoga), balance (strength training), and fat burning (aerobics). As it introduces the user to the aerobic exercises, it says, "With aerobics, you'll work on burning your body fat by doing light exercises for several minutes. Having said that, pushing yourself even just a little bit during your training can help strengthen your body and burn fat faster." It is clear that the dominant theme of the *Wii Fit* is not necessarily health or fitness; rather, weight loss is the goal of the user.

The trainers in the yoga and strength-training activities also reinforce the goal of weight loss. There are relatively few differences between the male and female trainers. Both provide encouragement when a user does well ("That's it! Feel those abs working"), and both are quick to criticize the user when he or she seems unbalanced ("You're a little shaky"). Both trainers refer to the user's weight: "Train every day and keep aiming for a toned waistline." However, it seems the male trainer is the only one to remark specifically on the user's body, regardless of the gender or BMI of the Mii: "While you're working out, visualize your ideal body."

Working to achieve a thin body enables patriarchal and postfeminist standards of beauty that dictate women's apparent "right" to work toward an unachievable body ideal.[24] Indeed, Jessica Francombe has observed in her analysis of the *We Cheer* game, a cheerleading game that plays on the Wii, "The impetus is with the young girl to mold her body, through makeovers and workout models, into the digital and internalized image of the ideal girl."[25] In the *Wii Fit*, sports games are ultimately appropriated as means to the end goals achieved through stringent self-maintenance and individual choices. Although the aerobic activities include running and boxing, they are not offered as competitive sports, but seen as ways of burning fat and losing weight. Sports like soccer, skiing, and snowboarding are adapted into balance exercises that help women achieve the ideal body. All of these sports are funneled into inspirational self-discipline, so that women can be thin and beautiful; thus, the goal of weight loss reinforces traditional displays of femininity.

As we have discussed, the position of the Wii in the home exacerbates the ideals of essential femininity in postfeminist terms. Beyond the sleek aesthetic of the console, influencing women to work out in the home ultimately reinforces a heteronormative gender-sex dichotomy; that is, women possess an essential femininity and should remain in the home to look after the family, whereas men are inherently masculine and should be the providers. Wives and mothers must maintain rigid beauty regimens that ultimately seem effortless. There is an unattainable ideal female aesthetic that is perpetuated through the *Wii Fit* under the disguise of fitness and posture. In addition, the sports activities that women are encouraged to play are also contained within the domestic sphere, and the aim of these activities is outside of the actual competition within the game, because the goal is continually fixated on body image. In other words, the real goal in the *Wii Fit* is not to win the games but to obtain beauty, a necessary commodity in the bigger game of catching and keeping a man.

The true intention of the *Wii Fit* is perhaps best demonstrated not by the console or the game package itself, or even the official ad campaign, but in a YouTube video that has become popularly know as the "*Wii Fit* Girl." The video begins with the camera focused on a young man's face, as he seems to be checking to make sure it is recording. The camera frame then pans to reveal a young woman facing the other direction, so that she is seen from behind – an important point, because she is wearing skimpy panties that ride over her buttocks. She begins the hula-hoop game on the *Wii Fit,* gyrating her hips and occasionally bending over to catch additional hoops. The camera pans back to the young man, who sticks out his tongue and wags it in a licentious leer. The video is posted on YouTube under different titles, but the one entitled "Why Every Guy Should Buy Their Girlfriend a *Wii Fit*" has had more than twelve million hits. The success of the video lends support to our argument about the unofficial and unspoken purpose of the *Wii Fit*. Fitness is merely the means to an end: having the kind of body that will attract a man. At least for the girl in this YouTube video, the *Wii Fit* works.

CONCLUSION

We began this chapter by noting that the console that would become the Wii received mixed reviews from the video game press. Undoubtedly, the *Wii Fit*'s commercial success is not in question, but as a means of improving personal health and fitness, the *Wii Fit* has had patchy success. In their analysis of the *Wii Fit,* Miyachi et al. found that a third of the program's activities can provide the necessary moderate-intensity exercise recommended by the American College of Sports Medicine and the American Heart Association. On the other hand, in another study published in the *Journal of Strength and Conditioning Research,* the health and fitness measure of eight families who used the *Wii Fit* for three months showed little change or improvement.[26] These two studies are not contradictory, but instead point to a logically consistent conclusion: the *Wii Fit* provides activities that may improve health, but success is not guaranteed.

If the postfeminist promise of empowerment is provided through consumer culture, then what happens to that promise when the consumer goods in question fail to deliver? In the case of the Wii and the *Wii Sports* and *Wii Fit* packages, that failure is both articulated by the console and visualized by the Mii. The women who use these devices are also invested with the responsibility of failure; it is never Wii's fault, because the women are to blame, and the console is quick to point that out, telling the overweight and obese to watch what they eat. With the incorporation of sports activities into the console, the Wii provides women with the paradoxical opportunity to win at boxing, but lose at being women. The *Wii Fit* falsely equates agency and empowerment with striving to attain idealized standards of beauty, which are unattainable for most women.

Women have always had the choice to play video games, yet Nintendo succeeded in marketing the *Wii Fit* to women as their game of choice. However, the goal of losing weight is hidden in language referring to posture and good health. Postfeminism suggests women strive to achieve the ideal body, yet must not make their efforts seem burdensome or obvious, a goal achieved by the simple act of playing an ordinary video game, an activity performed alone and in the home. Supposedly, these choices represent the authentic desires of the individual, but as we have seen, such desires are relegated to rigid gender roles and perpetuate patriarchal standards of femininity. Considering that television has been an important conduit for these standards, the inconspicuous physical design of the console and the interface belies something rather insidious: the *Wii Fit, Wii Sports,* and *Wii Fit Plus* are literally more channels for propagating postfeminist notions of femininity.

Although the Wii might be considered an important moment for women, in that it opened the video game market to them, and did so through a series of sport simulation games, the console obviously does not necessarily have women's interest in mind, aside from their interest in consuming. Instead, what the console reflects are the interests of women as understood by men. While pleasing Mom might have been the defining goal of the Wii design, it was overwhelmingly men who produced the actual design of the console. Furthermore, the way these men incorporated women's interest was to filter that interest through their own masculine understanding of women's roles as wives and mothers. It should come as no surprise that the way the male-dominated culture at Nintendo imagined the interests of women was in service to their own needs as fathers and husbands. In other words, pleasing women in the design of the Wii meant designing a console that helped women please men. Therefore, the Wii console, *Wii Sports,* and *Wii Fit* all reveal that the goals of postfeminism are ultimately the goals of prefeminism and that the choices that seem to be empowering women are very specific in their endgame: the female body as an object of sexual desire. And while the console may pretend to level the ground where sports are concerned, the game is already rigged in men's favor.

NOTES

1. Ryan Block, "Engadget and Joystiq's Live Coverage of Nintendo's Revolution E3 Event," *Engadget,* May 17, 2005, http://www.engadget.com/2005/05/17/engadget-amp-joystiqs-live-coverage-of-nintendos-revolution/; Mark Casamassina, "IGN-cube's Nintendo 'Revolution' FAQ," *IGN.com,* September 12, 2005, 2, http://www.ign.com/articles/2005/09/12/igncubes-nintendo-revolution-faq.

2. Rachel Rosmarin, "Nintendo's New Look," *Forbes,* February 7, 2006, 11, http://www.forbes.com/2006/02/07/xbox-ps3-revolution-cx_rr_0207nintendo.html.

3. W. Chan Kim and Renee Mauborgne, *Blue Ocean Strategy: How to Create*

Uncontested Market Space and Make the Competition Irrelevant (Boston: Harvard Business Review School Press, 2005), 11, 18.

4. Robert Alan Brookey, *Hollywood Gamers: Digital Convergence in the Film and Video Game Industries* (Bloomington: Indiana University Press, 2010).

5. Ibid.

6. Sales figures taken from VGChartz .com.

7. Jennifer Glos and Shari Goldin, "An Interview with Brenda Laurel (Purple Moon)," in *From Barbie to "Mortal Kombat": Gender and Computer Games*, edited by Justine Cassell and Henry Jenkins (Cambridge: MIT Press, 1998), 118–135.

8. Diane Negra, *What a Girl Wants? Fantasizing the Reclamation of Self in Postfeminism* (London: Routledge, 2009).

9. Ibid., 12.

10. Ibid.

11. Martin Roberts, "The Fashion Police: Governing the Self in *What Not to Wear*," in *Interrogating Postfeminism: Gender and the Politics of Popular Culture*, edited by Yvonne Tasker and Diane Negra (Durham, NC: Duke University Press, 2007), 227.

12. Negra, *What a Girl Wants*.

13. Yvonne Tasker and Diane Negra, "In Focus: Postfeminism and Contemporary Media Studies," *Cinema Journal* 44, no. 2 (2005): 107.

14. Negra, *What a Girl Wants*, 96.

15. Suzanne Leonard, "'I Hate My Job, I Hate Everybody Here': Adultery, Boredom, and the 'Working Girl' in Twenty-First-Century American Cinema," in *Interrogating Postfeminism*, edited by Tasker and Negra, 101.

16. Stephen Kline et al., *Digital Play: The Interaction of Technology, Culture, and Marketing* (Montreal: McGill-Queen's University Press, 2003), 121.

17. Ibid.

18. Osamu Inoue, *Nintendo Magic: Winning the Videogame Wars* (New York: Vertical Press, 2010), 38, 49, 66, 67.

19. Ibid., 49, 55.

20. Ibid., 46.

21. Paula W. Hinley, "Occupational Therapists Use Wii in Parkinson's Study," *MCG News*, April 7, 2008, https://my.mcg .edu/portal/page/portal/News/archive /2008/Occupational%20therapists% 20use%20Wii%20in%20Parkinson's% 20study; Jami Kinton, "With Wii Bit of Help, Rehabbers Doing Fine," *Mansfield (OH) News Journal*, December 27, 2008, http://www.mansfieldnewsjournal.com /article/20081227/NEWS01/812270307 /1002.

22. Adrienne Massanari, "Gendered Pleasures: The Wii, Embodiment, and Technological Desire," in *Social Exclusion, Power, and Video Game Play: New Research in Digital Media and Technology*, edited by David G. Embrick, Talmadge J. Wright, and Andras Lukacs (Plymouth, UK: Lexington Books, 2012), 115–138.

23. Andrew Prentice and Susan Jebb, "Beyond Body Mass Index," *Obesity Reviews* 2 (2001): 141–147. The reliance on BMI to illustrate the shape of the Mii is troublesome for a number of reasons. Most important, BMI is not a completely accurate measurement of fatness. Because a BMI is determined by overall weight and does not account for muscle or fat, some individuals with relatively little body fat may have a high BMI.

24. Ariel Levy, *Female Chauvinist Pigs: Women and the Rise of Raunch Culture* (New York: Free Press, 2005).

25. Jessica Francombe, "I Cheer, You Cheer, We Cheer: Physical Technologies and the Normalized Body," *Television and New Media* 11 (2010): 360.

26. Motohiko Miyachi et al., "METs in Adults While Playing Active Video Games: A Metabolic Chamber Study," *Medicine and Science in Sports and Exercise* 42 (2010): 1149–1153; Scott G. Owens et al., "Changes in Physical Activity and Fitness after 3 Months of Home Wii Fit Use," *Journal of Strength and Conditioning Research* 25 (2011): 3191–3197.

PART TWO

The Uses of Simulation

SIX

Avastars: The Encoding of Fame within Sport Digital Games

Steven Conway

LIONEL MESSI HAS DEVELOPED WELL DURING HIS TIME AS Surreal Madrid's star striker. He has an overall rating of 98, with an attack and shot accuracy of 99, dribble accuracy and dribble speed of 98, and explosive power of 97. Allied to this are eleven special abilities, such as "incisive run," "long-range drive," and "roulette skills" (this refers not to the casino game, but to the skill of pirouetting over a soccer ball to avoid an opponent's incoming challenge). He has evolved into the definitive "game changer," as we say in common managerial parlance. My other striker, the 1961 iteration of Brazil's Pelé, has a host of attributes in the high '90s with eighteen special abilities. The latest boot technology from Adidas's Predator range accentuates my strikers' already extraordinary proficiency; I chose the Predator for its high shot power and swerve ratings over the adiZero's high acceleration and top speed. After much careful tinkering with my squad's formation and tactics, I take to the pitch, prematch nerves building in the tunnel. Following a sublime performance, we have annihilated FC Barcelona 4–0 in the semifinal of the Champions League. The intense rivalry between the clubs is well documented by the press, and I am informed postmatch that Surreal Madrid's loyal fan base is distinctly pleased with the result; we are now an S (super) grade in popularity. This is particularly gratifying news for my scouts, who know that this rating may finally be the key to attracting Cristiano Ronaldo to put pen to paper for Surreal Madrid.

None of this, of course, happened outside of my television screen. Yet it did happen, albeit within the confines of my now priceless saved game of *Pro Evolution Soccer 2012*, where I exist as a kind of Hobbesian

leviathan: creator, owner, manager, agent, trainer, and frequent puppe-
teer of a galaxy of soccer superstars (or are some, such as Messi, now su-
pernova?). In regards to time spent engaged with the game, I am mostly
interpellated by the system to inhabit the subjectivity of an information
economy worker, tasked with the day-to-day management of numerous streams of data: the latest results of my training regime (carefully
designed to increase acceleration and top speed), squad contracts, sponsorship, market negotiations, formations and tactics, recently unlocked
technologies, and the careful balancing of the budget via expected income and outgoings.[1]

Such an intensely positivistic quantification of the athlete, sport
technology, and industry raises many questions concerning our understanding and consumption of sport. For the purposes of this chapter,
I will be examining the following interwoven phenomena. First, I will
consider the multifarious ways in which the concept of sport is discursively positioned within the video game. Second, I will explore how the
various equipment of sport (the body, the kit, the boot, the ball, and
so on) is re-presented and enacted within the digital interpretation as
hypervisible, hyperludic objects.[2] Third, I will discuss the profoundly
intertextual way in which the sports digital game constructs meaning,
and therefore makes sense to the consumer-player, via the cultural logic
of the information economy and the convergent strategy of the modern
media-sport complex. Fourth, I will illustrate how the digitized celebrity,
as supremely rationalized object, is a rhetorical maneuver by the forces
of production to further legitimize the celebrity status of players (and
therefore the legitimacy of the wider sport industry). To inflect Weber,
it is a move that translates the "charismatic authority" of the celebrity
into a "legal-rational authority" so dominant within, and conducive to, a
late-capitalist, bureaucratic society.[3] Charismatic authority, historically
associated with change (Lenin, Gandhi, Castro, and others), is coerced
by knowledge capitalism, which requires predictability and stasis: a "charisma" commodity, the celebrity, is produced. Fifth, I will examine how
this move toward legal rationality is the pinnacle of alienation vis-à-vis
reification; perhaps Marx's original German *entfremdung* (estrangement)
is germane to this context also, demonstrating not only the supreme
isolation from but also the frequent hostility toward these polysemic

identities the user so intimately knows through intertextual consumption and parasocial interaction.[4] To provide for the reader a consistent link between these sometimes disparate perspectives, I will focus mainly upon the representation of the soccer celebrity as the anchor through which we can identify broader sociocultural dynamics and tendencies.

In performing this analysis, I will utilize the postphenomenological probe of transparent and opaque technology following Andy Clark:

> In the case of such opaque technologies, we distinguish sharply and continuously between the user and the tool. The user's ongoing problem is to successfully deploy and control the tool. By contrast, once a technology is transparent, the conscious agent literally sees through the tool and directly confronts the real problem at hand. The accomplished writer, armed with pen and paper, usually pays no heed to the pen and paper tools while attempting to create an essay or a poem. . . . Sports equipment and musical instruments often fall into the same broad category.[5]

Genealogically, we may link this to the phenomenology of Heidegger and his discussion of *vorhanden* (present-at-hand) and *zuhanden* (ready-to-hand), and I will illustrate the utility of this probe at many levels of scale: the social, the cultural, the industrial, the informational, the ludic.[6] In speaking of transparency and opaqueness, I am articulating the relationship between the physical and digital versions of the sport and how certain concepts, rhetorical standpoints, and sporting technologies are rendered perceptible (often to the degree of palpability) through their variegated integration into the digital game apparatus as both symbolic and ludic agent.

Though this notion of technology's opacity may at first glance seem unique to the digital game interpretation of sport, in contrast to the "pure" physicality of the classic athlete, it is important to note that there is a long, often unremarked-upon history between sport and technology fraught with boundary negotiations, to say nothing of the links between modern sport and pharmacology.[7] My analysis will be supported, implicitly and explicitly, by the theoretical influences of Gramscian hegemony, Bruce Latour's actor-network theory (ANT), Marshall McLuhan's discussion of media and technology, and Graham Harman's object-oriented philosophy (which is to a large degree indebted to both Heidegger's and Latour's work).[8]

I will also use the ludological framework of hyper- and contraludicity. Briefly, ludicity is the baseline agency afforded to an object within the game (itself a network of objects) at that precise moment in time (the event). That is to say, the capacity for the object to have an effect upon the gamestate (defined as the cumulative total of the game's network of ludic objects). For example, following Harman's object-oriented philosophy, which corresponds well to the ontology of object-oriented programming, this object may be a ball, a chess piece, a board game tile, a hockey stick, the strength rating of an avatar, the numerous abilities of a user, and so on. This is not to say all objects have equal agency. As Ian Bogost points out, "All things equally exist, yet they do not exist equally," and generally it is the human user that maintains supreme ludic agency within the network of the digital game.[9]

Let us take, for example, a very basic game, such as Rock, Paper, Scissors: rock beats scissors, scissors beat paper, and paper beats rock. This game maintains consistent ludic agency between objects (assuming each user is equally competent). Yet if we were to introduce the object of "dynamite" for one user, which always beats the three eponymous objects, then that user would have access to hyperludicity, that is, an object far above and beyond the average ludicity. In a similar vein, one can identify *Super Mario*'s mushrooms, *Pac-Man*'s power pellets, *Call of Duty 4: Modern Warfare*'s airstrikes, the embodied skills and knowledges of professional athletes, and so on. Such objects provide an opportunity for the capable user to expedite their journey toward the game objective, whether that is achieving a high score, eliminating the opposition, or, as we will see for soccer, scoring goals. Contraludicity is the opposite, where an item – the game environment, the control scheme, an onlooker, or perhaps an opposing player – diminishes or disrupts the user's agency: broken equipment, bad weather, inverted control schemes, hustlers, streakers, and expert (or faulty) artificial intelligence constitute a broad selection of reference points.

THE SPORT DIGITAL GAME

The digital interpretation of physical sport is a highly intertextual product in the manner in which it borrows form and content from all kinds of artifacts: wargaming and its modern fantasy iterations (for example, *Dungeons and Dragons*), collectible card trading games, the graphical user-interface aesthetics of modern operating systems such as Microsoft Windows and Apple's OSX and iOS, television, tabloid newspapers, and even modern musical forms such as rap and hip-hop, which are particularly influential upon the audiovisual patina of the nascent "extreme" subgenre (for example, *NFL Street*).

While the aesthetics of the digital game medium reside in an immensely intertextual and paratextual universe, drawing in, and indeed constantly referencing, influences from other media and cultures, the sports digital game takes this to an even more intense degree in terms of production, as Baerg explains when discussing the creation of player statistics within such products (as indicated within the introduction, vital numerical information concerning the virtual athlete is incorporated, from height and weight to numeric assessments of dribbling speed, shooting power, and so on, normally rated between 0 and 99):

> Not only is the experience of playing the game intertextual (Crawford, 2008), but the very construction of the game exists as an intertextual process as well. Player ratings data provide one of the vital intertextual links between real world sport and its digital counterpart. The data fails to exist in its current form without the influence of other forms of mediation. According to Booth and Muller-Mohring, sports journalism, television broadcasts and sports talk radio shape the player rating construction process. Therefore, in building on Crawford's work, intertextuality plays as much of a role in the production of the game as it does in its consumption.[10]

Further, although the playing of a game is always an intertextual, and often intratextual, process, in terms of the sports video game there are a host of preexisting literacies and knowledges assumed by the developer on behalf of the player. One cannot understand the offside rule if they are previously unfamiliar with the sport; one can also not know where to deploy a player such as Liverpool's Steven Gerrard, and how to use his avatar, without either prior familiarity with his soccer career or through repeated playing of the digital game. The intertextuality of the sport video game, that is, the understanding one brings from previous experience with the sport-as-played and culture to his or her comprehension of it, is a synthesis of knowledge concerning a wealth of governing frameworks, embodied practices, media texts, and subcultures.

THE CELEBRITY . . . IN THEORY

Historian Daniel J. Boorstin mused famously upon celebrity, "The hero was distinguished by his achievement; the celebrity by his image or trademark. The hero created himself; the celebrity is created by the media. The hero is a big man [*sic*]; the celebrity is a big name."[11] Boorstin's rumination on the difference between hero and celebrity is balanced upon a precarious distinction: that hero and celebrity cannot be one and the same. It is at its core a critique of the validity of the celebrity's charismatic authority.[12] While, in the age of *Big Brother* and numerous other "reality" television shows, Boorstin's assertion is consonant with the public's view of the latest iteration of celebrity-product, "famous for being famous," it is a specious argument when applied to the category of sporting celebrity. This form of star is created through an often extensive list of achievements before the eventual image or trademark becomes associated with him or her, fueling the rise to celebrity: David Beckham is indeed distinguished by his image and trademark, yet his initial trademark was his free-kick technique, and thus also, first, an achievement.

In the realm of sport, in opposition to the "media-made" celebrity, one must first prosper in their chosen field before the limelight of fame is provided. As mentioned, Beckham created his trademark first on the soccer pitch, Tiger Woods on the golf course, Wayne Gretzky on the ice rink.[13] The celebrity status that followed was a result of these achievements, reward for their extraordinary ability within their discipline.

Chris Rojek makes the somewhat useful distinction between Achieved Celebrity and Attributed Celebrity: the former the well-tread ground of the public intellectual, the best-selling writer, the rock star, and, of course, the sporting celebrity, the latter that upon which Boorstin's vitriol is directed and which was articulated in Gramscian terms by Richard Dyer as "owing their existence solely to the machinery of their production." This kind of manipulation theory, predicated upon belief in the hegemonic forces of capitalism churning out the mass-consumable, prefabricated celebrity that the public then swallows whole and without question, is obviously challenged by the agency of the audience. Cultural theorists such as Stuart Hall, John Fiske, and Henry Jenkins have been quick to dismiss such top-down views of cultural production, illustrating well the polysemy of even the most one-dimensional caricature: simply put in regards to celebrity, manipulation theory "is built on an implied universal value system that confidently claims to distinguish substance from vacancy, whereas fans will readily attest that one person's attributed celebrity is another person's achieved celebrity."[14]

Focusing upon the sporting icon, Barry Smart offers a compelling analysis of the modern sports star as a culture industry phenomenon, sketching the complex relation between sporting heroics, celebrity, and commodification. Smart traces the contemporary integration of sports hero and celebrity as a consequence of the early twentieth century's emergence of marketing and public relations as dominant sociocultural forces alongside television, film, and the Hollywood star system: "Sport, as a whole, including the profiles of individual athletes and players, has been transformed by the impact of processes of commercialization and media amplification into a spectacle. . . . [A]s communications media have developed the profile of the sporting figure has been raised and extended, if not radically transformed." The everyday embeddedness of this kind of charismatic figure within society, filling our magazines, television shows, and radio broadcasts, is in direct opposition to Weber's original definition: "Charismatic authority is thus specifically outside the realm of everyday routine and the profane sphere. In this respect, it is sharply opposed both to rational, and particularly bureaucratic, author- ity. . . . Bureaucratic authority is specifically rational in the sense of being bound to intellectually analyzable rules; while charismatic authority is specifically irrational in the sense of being foreign to all rules." As we will see, this incorporation of the charismatic figure into the everyday is part of a strategy of manufacturing consent, in a Gramscian sense, by various hegemonic institutions, alongside and inseparable from the insatiable appetite of capitalism to commodify that which reaches critical mass in terms of cultural capital.[15]

Garry Whannel has provided perhaps the most insightful, and cer- tainly the most focused, analysis, detailing the ways in which the mod- ern sport star is, again, a charismatic, almost mythic figure, or at least an anthropomorphic gateway into this mythic world of heroes and villains, gods and demons. Sport has long existed for the fan as a parallel universe, existing outside the space and time of ordinary life, a magic circle, as Huizinga first described in his well-worn quote: "temporary worlds within the ordinary world, dedicated to the performance of an act apart." This is not to say that this temporary world is hermetically sealed (as Crawford, among many others, has pointed out), but that on the level of affect and social formation, it retains a level of impact and form of organization that is distinctly separate from the mundane, the boring, and the everyday.[16]

Indeed, it is more accurate to say that these temporary worlds, in which sport, leisure, and gaming reside, can be seen to accentuate Huizinga's "ordinary world." As Whannel discusses, the modern sport star has a broad cultural visibility that is firmly embedded within everyday media consumption: sport is an essential facet of cultural hegemony, not apart from it.[17] Therefore, the cultural and economic roles of sport are never truly apart from one another: the legitimating process of sport – for example, its professionalization, formation as an industry, and expanding commodification – is dependent first and foremost upon a stable economic base.

The sport celebrity, as a focal point of this legitimation, is therefore forced to carry much ideological baggage, some of which they may be reticent to take upon themselves. Case in point is NBA star LeBron James, whose famous "What Should I Do?" Nike advertisement simultaneously parodies and rejects the hegemonic conception of all-American "sporting hero" so often forced upon the public persona of those who become household names. In doing so, Nike is performing the classical move of incorporation as identified by Gramsci, constructing its brand identity by leveraging that which is seen as resistant and transgressive within hegemonic sport culture, paradoxically attempting to integrate such transgression into the dominant paradigm through acknowledgment and admission.[18] Instead of becoming a potentially dangerous, disruptive agent to the public persona of the National Basketball Association and sport more generally, LeBron is re-presented and reincorporated as an almost parodic "villain" caricature, a cartoonish mischief maker, and therefore preemptively neutered, rendered harmless in regards to the NBA's image.

A STAR IS BORN

Daniel Kromand, in his paper on avatar categorization, offers a useful typology for avatars. Kromand employs four defining aspects across two axes (open-closed and central-acentral) to create four categories of avatar.[19] "Open" or "closed" defines the avatar's creation: an open avatar allows the user to manipulate appearance, ability, and personality traits, while closed avatars have such features preinstalled. For example, Lara Croft from *Tomb Raider* is a closed avatar, with her own appearance, skills, and motivation, while an open avatar would be one created by the user for Blizzard's *World of Warcraft*. A central avatar is under the user's direct control, such as in *Sonic the Hedgehog,* while an acentral avatar is indirectly controlled through ancillary guidance and manipulation of the environment, such as in EA's *The Sims.*

The central-closed avatar of the soccer genre is not just preinstalled by the developer; it is preinstalled by the wider sports-media complex and community through continuous coverage of the soccer players' lives, both professional and private, and to a fan a downturn in form, or even a new hairstyle, can make the previously authentic televisual simulation model now appear markedly inaccurate. In previous iterations of EA Sports' *FIFA* and Konami's *Pro Evolution Soccer* series, David Beckham's haircut and dead-ball animations were continuing obsessions for the producers; memorably, one of the Microsoft "Xbox live" updates for *Pro Evolution Soccer 5* even included an update to Beckham's hairstyle, and the way he strikes the ball when taking a free kick is included as part of the game's visual spectacle. Cristiano Ronaldo's noted free-kick technique has been included in the latest *Pro Evolution Soccer* and *FIFA* products, and minor details such as these are of the utmost importance to the franchises' fans, so much so that they are regularly monitored and commented upon by the gaming press in reviews. The mercurial nature of the professional soccer media environment has culminated in the institutionalization of fan modification; certain games, such as the *Pro Evolution Soccer* series, incorporate a powerful editor into the final product. Thus, the consumer's dynamic, highly intertextual consumption of modern sport is acknowledged and, further, endorsed by the producers.

The user can modify any current player's statistics, from their special abilities and dribbling speed to aesthetic choices such as free-kick posture, running style, and boot design. Depending upon the level and amount of edits, one fan's version of the video game may be viewed as more or less authentic than another, always in relation to the wider sports-media complex within which the concept of "authenticity" is contested and constantly reforged. The act of increasing one superstar's representational or ludic opacity, whether that is their new signature haircut or shot power, is never an isolated phenomenon but always bound up in networks of intertextual, transmedial tensions.

THE HYPERLUDIC CELEBRITY

For the sports digital game, the baptismal of fame often manifests as both representational adornment and ludic catalyst. For example, the celebrity is clearly identifiable within these products through visual flourishes indicative of a kind of tribal fetishism; within the *FIFA, John Madden Football* (later renamed *Madden NFL*), *NBA 2K*, and *Pro Evolution Soccer* series, the celebrity is commonly anointed with unique hairstyles, wristbands, gloves, and footwear, all of which may carry ludic enhancements. Rare animations are implemented that further accentuate the idiosyncrasy of the celebrity so they may stand apart from their peers: Michael Jordan, Kevin Durant, Derrick Rose, et al., maintain a selection of "signature skills" and dunks in *NBA 2K13;* Lionel Messi and Cristiano Ronaldo perform the extremely rare *elastico* skill in the *Pro Evolution Soccer* series (which involves the flicking of the ball first outward and then rapidly back inward with the same foot in an attempt to confuse the defender).

To illustrate some examples from the digital interpretation of soccer, specifically within the televisual genre of the sport digital game (that is, those that remediate the televisual image of sport), such as EA Sports' *Madden* and *FIFA* franchises, the celebrity's hyperludicity is evident in various ways.[20] First, as mentioned in the introduction, the avatar's "ability settings" have been adjusted to exceptional levels. Ability settings are the basic capabilities of the avatar; for example, *Pro Evolution Soccer 2012* maintains a list of twenty-nine abilities, from *attack* and *defense* (determining the avatar's artificial intelligence, or AI, apropos situational awareness for attacking and defensive maneuvers) to *weak foot frequency* (the likelihood of the player shooting with his weaker foot) and *acceleration* (the higher the value, the quicker the avatar reaches the *top speed* rating). Massaging these basic statistics are a further possible twenty-three *special abilities,* such as *one-on-one shot* (making it easier to score goals when one-on-one with the goalkeeper), *outside of the boot* (increases the player's accuracy when using the outside of his foot), and *long throw* (the avatar is capable of extraordinarily long throw-ins).

A celebrity player such as Manchester United's Wayne Rooney has very high ability settings combined with many special abilities (empirically, it is rare to discover a player with more than three special abilities, while Rooney has nine of the twenty-three). This means that not only is the avatar of Wayne Rooney gifted above the majority of his peers through simple abilities such as *aggression* and *stamina,* but in terms of special abilities the effect he can have upon the gamestate is enormous compared to an unknown player, who in comparative terms is ludically disabled.

For example, Rooney has the *shooting from distance* special ability, meaning it is easier to shoot on target from more than eighteen yards, while a player without the ability would be more likely to shoot the ball aimlessly over the bar. He also has the *sliding tackle* special ability, described by the in-game tool tip as being "able to slide-tackle well with less chance of committing a foul." He is therefore, through this special ability, elevated above the rules of the sport: whereas other anonymous players would be penalized for the slide tackle, Rooney is unlikely to suffer the consequence. This is an expansion of hyperludicity to a point where the *agon* principle of soccer, the notion of equal contestation, is nullified, and therefore the hyperludic becomes *hypo*ludic: the winner of the challenge is a bygone conclusion, and play is to a large degree nullified.[21]

Tied to this are certain themes that emerge through the analysis of these attributes, in terms of the long-standing division between "British" and "foreign" player qualities, as explained by Archetti: "Two styles were thus contrasted, the *british* and the *criollo.* 'Britishness' was identified with being phlegmatic, disciplined, methodical, and concentrated on elements of the collective, of force and of physical power. These virtues help to create a repetitive style, similar to that of a 'machine.' The *criollo*, due to the Latin influence, was exactly the opposite: restless, individualistic, undisciplined, based on personal effort, agile and skilful."[22] Thus, there is a tendency in soccer digital games for the British celebrity, as symbolic of the British style of play, to have high *stamina, body balance* (the likelihood of them winning a shoulder-to-shoulder confrontation), and *team work,* with associated special abilities of strength and physicality (such as the *sliding tackle* ability), whereas the foreign celebrities have characteristics reflecting their individuality and skillfulness.

To compare, if we look at Rooney's former Manchester United teammate Cristiano Ronaldo, a Portuguese player famous for his dribbling skills and sometimes ostentatious behavior, while his *stamina* and *teamwork* ability are conspicuously low, his *dribble accuracy, dribble speed, agility,* and *technique* ratings are almost at maximum (the ratings are illustrated as out of a possible 99, with Ronaldo above 95 in the latter four, while he is rated below 80 in the first two). He is given six special abilities, with three focusing on his "personal effort," as Archetti would say; *dribbling* and *tactical dribble,* the first explained as such in-game: "The player (when controlled by artificial intelligence) favors dribbling." The *outside of the boot* special ability is also given (the player can shoot effectively in unorthodox positions). All of these enhance his individual prowess and general *criollo* style.[23]

Considering these overall concessions afforded to the celebrity, the supposed preternatural talent of the sporting superstar, mythologized through the vortextuality of the modern media-sport complex (obsessively referencing the "golden moments" of their careers, their extraordinary skills, and personality), it is clear how the often unexplainable, nebulous, and fleeting concept of "sporting genius" is solidified through the discursive power of scientific naturalism: everything can, ultimately, be measured, explained, and assigned a value.[24]

A significant side effect of this ludic reification of the sporting celebrity is to advance the allure of the "open product," as Masuyama describes:

> The most significant aspect of the Pokemon concept is the fact that it is not a [closed] product. Take film: we see a beginning, an end, and finally, the credits rolling. That is a "closed" product. The Game Boy title Pokemon also has a

beginning and an end, and even credits.... [But] that hardly means the game is over. At the end, in order to complete the Monster Encyclopedia, the player sets out once again . . . in search of Pokemon. Also, a player may use a cable to trade Pokemon with other players.[25]

The existence of these rare celebrity avatars creates for the player a form of metagame following the aesthetics of collectible card games, whereby they attempt to collect as many uniquely gifted star players as possible, to compare and contrast with fellow players.[26] As detailed, *Pro Evolution Soccer* encourages this through the unlocking of "legendary" players such as Brazilian Pelé and Argentinean Diego Maradona by meeting certain criteria (generally winning certain leagues and cups) in an exhaustive cycle of acquisition. EA Sports' *FIFA* has taken this to its logical conclusion in its latest versions, where a game mode, "FIFA Ultimate Team," centers upon building a squad by collecting players via a collectible card dynamic; users can pay real-world money to buy card packs or earn these packs by winning leagues and tournaments. The cards within these packs are highly randomized, a variable outcome design in the tradition of B. F. Skinner's famous box.

In approaching not only the celebrity but also more broadly the athlete as reducible to an aggregate of numerical ratings within these products, the often transparent operation of the human body is condensed into a supremely rationalized, opaque technology awaiting deployment. This is a narrow caricature of the athlete, existing as nothing but profession; it is in essence the reification of the body, symptomatic of the "big data" trend within knowledge capitalism, and therefore propagates the "distinctive form of 'rationality' which emphasizes abstract, quantitative calculability to the exclusion of other forms of human sensibility."[27]

THE EUPHEMISTIC CELEBRITY

Of course, while minor details of their appearance and abilities are given the utmost care and attention by the developers and press, the negative qualities associated with the celebrity in question are always conspicuous by their absence; the postmodern "antihero" is not yet part of the sport digital game's vocabulary. Luis Suarez has no predilection for diving and cheating, a notion widely held in the media to be one of his main

traits; Diego Maradona, once unlocked from the legendary set of players, does not have the capacity to perform *La Mano de Dios* (the hand
of God) to score a goal; Roy Keane is not prone to rash challenges or to
being booked more often than any other avatar because of his volatile
reputation; and so on.

As noted, the capability of the "avastar" within the video game is
enhanced so that they may prove themselves worthy of the "godlike"
mantle of celebrity; they must meet expectations preinstalled by the
sports-media complex vis-à-vis their extraordinary prowess.[28] In this
way, the digital game representation of the sporting hero runs counter
to the traditional relation between professional player and fame, as Ohl
describes: "Sport, more than all other domains, demands that, whoever
you are, even if you are a world champion, you must train and fight to
stay at the top. From a symbolic point of view, sport is the antithesis of
the culture of inheritance and privilege."[29]

The digital sports star, then, is the epitome of "inheritance and privilege" so often countered by public discourse on sport. If a common narrative within sport stardom is rags to riches, something from nothing,
then the digital star is born with a silver spoon in its mouth. Instead of
being depicted in line with the dominant view, as a person who trains
the longest, works the hardest, goes the extra mile (quite literally) to
become the best, the digital superstar is always already the best, seemingly without effort.

Yet such representation is a process of legitimation endorsed and
indeed required by the broader media landscape and culture industry
at large, as Rojek further ruminates: "Mass-media representation is the
key principle in the formation of celebrity culture.... [C]elebrities often
seem magical or superhuman. However, that is because their presence
in the public eye is comprehensively staged."[30] Such staging of this charismatic force reaches its zenith within the realm of digital games, as the
medium's aesthetic is governed by verbs rather than nouns and adjec-
tives: as a channel for action first and foremost, the celebrity must not
only portray but also enact arête in video games. As Socrates was reputed
to say, "Fame is the perfume of heroic deeds," and deeds are the digital
avatar's raison d'être. We are therefore confronted by a paradoxical state
of affairs within the digital interpretation of sport: without hyperludic-

ity, the celebrity avatar would not be famous, and without fame, the celebrity avatar would not be hyperludic. The extraordinary ability and charisma of the sport celebrity as perpetuated by the cultural industries are thus translated into a model comprehensible within the information economy, a form of legal-rational authority defined and encoded through finite rules, categories, and the ensured solidity of numbers.[31]

CONCLUSION

In its totality, the soccer video game wipes the semiotic slate clean and reconstructs a utopian presentation of the sport aligned with the mathematical precision and neutrality of the machine, with the potential for any political, social, or cultural subversion of the sport (as sometimes practiced by the players and fans themselves) nullified. A euphemistic interpretation is evident, yet as ever this is due to the sports digital game coming into being through a confluence of objects: from newspaper coverage of Wayne Rooney's temperament to statistics supplied by sports data companies to our most entrenched sociocultural values. By transforming the celebrity into a hyperludic game piece primed for heroic acts, the developers propagate a form of cult worship synchronic with mass-media production.[32] The celebrity is the charismatic authority of capitalism par excellence: a simulacral authority that is intimately tied up with the dynamics of the market, wielding, instead of the power to direct change and upheaval, the power to direct forces of acquisition and consumption.[33]

In portraying these celebrity soccer players as godlike, preternatural beings worthy of special concession, the digital game is simply perpetuating a dominant feature of postmodern media coverage necessitated by market forces. The potential polysemy of the celebrity, and therefore their autonomy as a living, breathing signifier capable of multiple, complex signs, is eclipsed by the monosemy so often demanded by the culture industry apparatus.[34] The sporting celebrity is defined as a commodity by his or her predictable one-dimensionality: Lionel Messi must be the humble rags-to-riches playmaker extraordinaire, Wayne Gretzky the something-from-nothing genius of ice hockey, LeBron James the outspoken, controversial talent.

The process of reification creates a golem, and for the sporting celebrity this is a golem defined by both intra- and extradiegetic processes of rationalization and quantification, whether this is *shot power* within the digital game or exchange value within the stock market. A recent example is provided in the case of Tiger Woods, whose public persona must maintain the image of the wholesome, all-American athlete, not the adulterous lecher he was found to be backstage; literally billions of dollars rest upon such an image, as Knittel and Stango found: "We estimate that shareholders of Tiger Woods' sponsors lost $5–12 billion after his car accident, relative to shareholders of firms that Mr. Woods does not endorse.[35] The losses are both substantial and widespread across many millions of shareholders." Luckily for Woods, as we have seen, the sporting celebrity's fame is a fame constituted of verbs, of athletic performance. Thus, the twin relationship of athlete and celebrity means that the narrative of redemption is always possible; the athlete can always become the celebrity once more, as the president of EA Sports, Peter Moore, responded to Woods's indiscretions: "By his own admission, he's made some mistakes off the course. But regardless of what's happening in his personal life, and regardless of his decision to take a personal leave from the sport, Tiger Woods is still one of the greatest athletes in history."[36] With Woods's power, workability, and spin in the '90s in *Tiger Woods PGA Tour 13*, it's hard to disagree.

LUDOGRAPHY

Blizzard Entertainment, *World of Warcraft* (Irvine, CA: Blizzard Entertainment, 2004)

Core Design, *Tomb Raider* (London: Eidos Interactive, 1996)

EA Sports, *FIFA Football* series (Redwood City, CA: Electronic Arts, 1993–present)

EA Tiburon, *NFL Street* (Redwood City, CA: Electronic Arts, 2004)

Electronic Arts, *Madden NFL* series (Redwood City, CA: Electronic Arts, 1989–present)

Gygax, G., and D. Arneson, *Dungeons and Dragons* (Lake Geneva, WI: Tactical Studies Rules, 1974)

Infinity Ward, *Call of Duty 4: Modern Warfare* (Santa Monica, CA: Activision, 2007)

Konami, *Pro Evolution Soccer* series (Tokyo: Konami, 1996–present)

Maxis, *The Sims* (Redwood City, CA: Electronic Arts, 2000)

Namco, *Pac-Man* (Tokyo: Namco Midway, 1980)

Nintendo Creative Department, *Super Mario Bros.* (Kyoto: Nintendo, 1985)

Sonic Team, *Sonic the Hedgehog* (Tokyo: Sega, 1991)

Visual Concepts, *NBA 2K* series (New York: 2K Sports, 2005–present)

NOTES

1. Louis Althusser, *Lenin and Philosophy, and Other Essays* (New York: Monthly Review Press, 2001).

2. Steven Conway, "Hyper-ludicity, Contra-ludicity, and the Digital Game," *Eludamos: Journal for Computer Game Culture* 4, no. 2 (2010a): 135, http://www.eludamos.org/index.php/eludamos/article/view/vol4no2-2.

3. Max Weber, *The Theory of Social and Economic Organization* (New York: Free Press, 1997).

4. Val Burris, "Reification: A Marxist Perspective," *California Sociologist* 10, no. 1 (1988): 22–43; Donald Horton and Richard R. Wohl, "Mass Communication and Para-social Interaction: Observations on Intimacy at a Distance," *Particip@tions* 3, no. 1 (2006).

5. Andy Clark, *Natural-Born Cyborgs: Minds, Technologies, and the Future of Human Intelligence* (Cary, NC: Oxford University Press, 2004), 37–38.

6. Martin Heidegger, *Being and Time,* translated by John Macquarrie and Edward Robinson (New York: Harper Perennial Modern Thought, 2008).

7. Sigmund Loland, "Technology in Sport: Three Ideal-Typical Views and Their Implications," *European Journal of Sport Science* (2002), http://www.tandfonline.com/doi/abs/10.1080/17461390200072105; Ivan Waddington, *Sport, Health, and Drugs: A Critical Sociological Perspective* (London: Routledge, 2000).

8. Antonio Gramsci, *Prison Notebooks* (New York: Columbia University Press,

1992); Bruno Latour, *Reassembling the Social: An Introduction to Actor-Network Theory* (Oxford: Oxford University Press, 2005); Marshall McLuhan, *Understanding Media: The Extensions of Man* (New York: Mentor, 1964); Graham Harman, *The Quadruple Object* (Winchester: Zero Books, 2011).

9. Conway, "Hyper-ludicity, Contra-ludicity, and the Digital Game"; Harman, *The Quadruple Object;* Ian Bogost, *Alien Phenomenology; or, What It's Like to Be a Thing* (Minneapolis: University of Minnesota Press, 2012), 11.

10. Steven E. Jones, *The Meanings of Video Games: Gaming and Textual Strategies* (Abingdon: Routledge, 2008); Gavin Stewart, "Paratexts: What Lessons Can We Learn from Inanimate Alice and Corporate ARGs," paper presented at "Under the Mask: Perspectives on the Gamer" (Luton, Bedfordshire, 2009); Andrew Baerg, "Genetic Metaphors, the Body, and Digital Basketball: Dynamic DNA in *NBA Live '09,*" *Journal of Communication Studies,* no. 2 (2010): 8–9.

11. Daniel Boorstin, *The Image: A Guide to Pseudo-events in America* (New York: Random House, 1992).

12. Weber, *Theory of Social and Economic Organization.*

13. Steven J. Jackson, "Gretzky Nation: Canada, Crisis, and Americanization," in *Sport Stars: The Cultural Politics of Sporting Celebrity,* edited by David L. Andrews and Steven J. Jackson (London: Routledge, 2001).

14. Chris Rojek, *Celebrity* (London: Reaktion Books, 2004); Richard Dyer, *Stars* (London: British Film Institute, 1998), 13; John Fiske, *Understanding Popular Culture* (London: Routledge, 1989); John Fiske, *Reading the Popular* (London: Routledge, 1989); Henry Jenkins, *Textual Poachers: Television Fans and Participatory Culture* (London: Routledge, 1992); Lorraine York, "Star Turn: The Challenges of Theorizing Celebrity Agency," *Journal of Popular Culture* 46, no. 6 (2013): 1332.

15. Barry Smart, *The Sport Star: Modern Sport and the Cultural Economy of Sporting Celebrity* (London: Sage, 2005), 17; Weber, *Theory of Social and Economic Organization,* 361; Gramsci, *Prison Notebooks;* Pierre Bourdieu, *Distinction: A Social Critique of the Judgement of Taste* (London: Routledge, 1984).

16. Garry Whannel, *Media Sports Stars: Masculinities and Moralities* (London: Routledge, 2002); Johan Huizinga, *Homo Ludens: A Study of the Play-Element in Culture* (London: Beacon Press, 1971), 10; Garry Crawford, *Video Gamers* (London: Routledge, 2011).

17. Garry Whannel, "Punishment, Redemption, and Celebration in the Popular Press: The Case of David Beckham," in *Sport Stars,* edited by Andrews and Jackson; Gramsci, *Prison Notebooks.*

18. Gramsci, *Prison Notebooks.*

19. Daniel Kromand, "Avatar Categorization," in *Proceedings from DiGRA 2007: Situated Play* (Tokyo: DiGRA, 2007), 400–406.

20. Jay David Bolter and Richard Grusin, *Remediation: Understanding New Media* (Cambridge: MIT Press, 1999).

21. Roger Caillois, *Man, Play, and Games* (Urbana: University of Illinois Press, 2001); Steven Conway, "We Used to Win, We Used to Lose, We Used to Play: Simulacra, Hypo-ludicity, and the Lost Art of Losing," *Westminster Papers in Communication and Culture* 9, no. 1 (2012): 27–46.

22. Eduardo P. Archetti, "The Spectacle of a Heroic Life: The Case of Diego Maradona," in *Sport Stars,* edited by Andrews and Jackson.

23. Ibid.

24. Garry Whannel, "News, Celebrity, and Vortextuality: A Study of the Media Coverage of the Michael Jackson Verdict," *Cultural Politics* 6, no. 1 (2010): 65–84; Teresa Lacerda and Stephen Mumford, "The Genius in Art and in Sport: A Contribution to the Investigation of Aesthetics of Sport," *Journal of the Philosophy of Sport* 37, no. 2 (2010): 182–193, http://www.tandf online.com/doi/abs/10.1080/00948705 .2010.9714775.

25. Meguro Masuyama, "Pokemon as Japanese Culture?," in *Game On: The History and Culture of Videogames,* edited by L. King (London: Laurence King, 2002), 41.

26. Richard Garfield, "Metagames," in *Horsemen of the Apocalypse: Essays on Roleplaying,* edited by Jim Dietz (Sigel, IL: Jolly Rogers Games, 2000).

27. Val Burris, "Reification: A Marxist Perspective," *California Sociologist* 10, no. 1 (1988): 34.

28. Rojek, *Celebrity.*

29. Fabien Ohl, "Staging Identity through Consumption: Exploring the Social Uses of Sporting Goods," in *Sport, Culture, and Advertising: Identities, Commodities, and the Politics of Representation,* edited by David L. Andrews and Steven J. Jackson (London: Routledge, 2005), 255.

30. Rojek, *Celebrity,* 13.

31. Weber, *Theory of Social and Economic Organization.*

32. John Maltby et al., "Thou Shalt Worship No Other Gods – Unless They Are Celebrities: The Relationship between

Celebrity Worship and Religious Orientation," *Personality and Individual Differences* 32, no. 7 (2002): 1157–1172.

33. Jean Baudrillard, *Simulacra and Simulation* (Ann Arbor: University of Michigan Press, 2004).

34. John Storey, *Cultural Theory and Popular Culture: An Introduction,* 5th ed. (Harlow: Pearson Education, 2009).

35. Erving Goffman, *Relations in Public: Microstudies of the Public Order* (New York: Basic Books, 1971).

36. Christopher R. Knittel and Victor Stango, "Shareholder Value Destruction Following the Tiger Woods Scandal," 2010, 6, http://faculty.gsm.ucdavis.edu /~vstango/tiger004.pdf; Ethan Sacks, "EA Sports Announces It Will Keep Scandal-Plagued Tiger Woods' Name on PGA Tour Online Video Game," *New York Daily News,* January 5, 2010, http://www.nydaily news.com/entertainment/gossip/ea -sports-announces-scandal-plagued-tiger -woods-pga-tour-online-video-game -article-1.460079.

SEVEN

Keeping It Real: Sports Video Game Advertising and the Fan-Consumer

Cory Hillman and Michael L. Butterworth

IN THE UNITED STATES, FEW, IF ANY, CULTURAL ACTIVITIES, products, or experiences are immune to the often unrestrained hands of commercialism, marketing, and advertising in the ambitious and overzealous pursuit of audiences and consumers. Sports are especially subject to these conditions, evidenced by the following examples: advertisers spent approximately $10.9 billion on national sports broadcasts between the final quarter of 2010 through September 2011; NBC paid the International Olympic Committee $2.2 billion to broadcast the 2010 and 2012 Winter and Summer Olympics; CBS and Turner Broadcasting agreed to pay nearly $11 billion to the NCAA for the rights to the annual men's college basketball tournament. Divisional realignment in college football has also been stimulated by the desire to create "megaconferences" in the chase for lucrative television packages with major networks, and the NCAA's decision to determine its national champion of college football's Football Bowl Subdivision with a four-team playoff beginning in 2014 came with estimates that the tournament could be worth as much as $6 billion.[1] Meanwhile, fans spent $3.2 billion on Major League Baseball (MLB) team merchandise in 2011, marking an 8.1 percent increase from the previous season, and the typical NFL fan spends approximately $60 on apparel, snacks, and other merchandise during the week of the Super Bowl.[2]

Clearly, commercialism has fundamentally altered the structure and organization of professional and big-time college sports, inviting a fair share of criticism and resentment. What is less clear is how the burgeoning commercialism of sports repositions and reshapes fan identity as a consumer transaction while manufacturing consent toward the strengthening and noticeable alliance being forged between sports, advertising, and capitalism.

This chapter focuses on sports video games and how they provide sports leagues and organizations with the opportunity to brand themselves within preferred and selective frames of representation and engagement, through rhetorical claims to realism that titles such as *Madden NFL* and *MLB: The Show* promise to deliver. In particular, in-game advertisements aim, among other things, to naturalize the growing commercialism of sports, making it appear as real as live sports themselves. In the process, sports video games function discursively with other cultural narratives to constitute a particular audience of *fan-consumers* in which fandom becomes naturally associated with both the obligation and the willingness to consume in performance of that identity. As notions of contemporary democracy are commonly conflated with consumption, the status of sports fans becomes subject to the same commercial mechanisms found elsewhere in culture that reduce the expression of identity to a consumer obligation.

In her study of mass consumption in the years after World War II, Lizabeth Cohen contends that purchasing habits in the United States were once seen as productive contributions to a healthy democratic society. Operating as a "citizen consumer," the average American could "be defined as one who consumed responsibly with the public interest in mind." Cohen goes on to map the transformation to the "purchaser consumer," a depoliticized agent hailed by the "consumerization of the Republic, where consumers/citizens . . . increasingly related to government itself as shoppers in a marketplace."[3] The consequences of the shift have been severe for democracy, as this discursive transformation "urges Americans to enact their citizenship through consumption," thereby transforming consumption from a contributing part to the larger whole of citizenship to the very definition of citizenship itself.[4] Thus, citizenship for Americans becomes less about public engagement with ideas, issues, and arguments and more about private acts of "choice" in the free market (voting included).

We would counter this notion of consumption as citizenship through an endorsement of what Robert Asen calls a "discourse theory of citizenship." From such a perspective, citizenship is a "mode of public engagement" that, while "potentially unruly," has the capacity to "open up unforeseen possibilities." In other words, it conceives of citizenship through modes of public interaction. At first glance, it might appear that sport offers a suitable public context for facilitating such interaction. After all, sport assembles people from across various communities and identities, bringing them together to share in the common cause of supporting their favored teams and players. Herein lies the essence of sport's democratic mythology, expressed a century ago in *Baseball Magazine*'s conclusion that baseball was an enactment of Thomas Jefferson's hopes for democracy, because the stands contained fans from all walks of life, "banker and bricklayer, lawyer and common laborer."[5] In more contemporary discourse, sport continues to be celebrated for its purported equalizing effects, whether it is among fans in the stands or by virtue of its apparent equal opportunities for upward mobility afforded to participants within the games themselves.[6]

For any number of reasons, most of which are beyond the scope of this chapter, sport has failed to actualize the democratic culture it so often is assumed to exemplify. Beyond the notable incidents of classism, racism, sexism, and homophobia, and beyond the hypernationalist ideology that is especially prevalent in the United States, sport has also been commodified in ways that increasingly marginalize "average" citizens in favor of affluent consumers who see in sport an identity achieved through acts of consumption. More than the billions of dollars at stake in free-agent contracts and television deals, fans are subjected to innumerable efforts to collapse the distinction between sport itself and its commercial interests.[7] Accordingly, as Lawrence Wenner summarizes, "consumer culture infuses our sporting identities and explains much about how our communities are approached and how we think of, and talk amongst, ourselves."[8]

Taking our cue from rhetorical scholar Maurice Charland, we suggest in this chapter that commercialized sport's ability to shape how we think and talk reflects its "constitutive nature" – that is, sport "calls its audience into being" through its ideological commitments.[9] From this view, the increasing commodification of sport further damages a democratic culture wherein citizens already understand themselves primarily as consumers. In the case of sports video game advertising, these relatively new practices draw upon previously established constitutive rhetorics that have shaped audience expectations about the nature of both sport and democratic citizenship. The remainder of this chapter, therefore, addresses the ideological implications of this rapidly growing sports-media convergence.

IT'S ALL ABOUT IMAGE . . . MANAGEMENT

In 2004 ESPN announced that it was canceling the highly rated but controversial series *Playmakers* over concerns that it would jeopardize its opportunity to renew its existing $600 million television contract with the NFL, set to expire in 2006. While *Playmakers* was about a fictional professional football league, the NFL assumed that the show's portrayal of spousal abuse, drug usage, and homosexuality might reflect negatively on its own brand of professional football. ESPN executive vice president Mark Shapiro stated that "the major issue for the league was that [*Playmakers*] lacked redeeming characters and redeeming qualities." One journalist contended that the show "went beyond the bounds of believability," while Gatorade dropped its advertising during the show.[10] Even NFL commissioner Paul Tagliabue approached Disney chief executive officer Michael Eisner with several complaints related to how *Playmakers* was tarnishing the institutional reputation of the league. According to Shapiro, "They just didn't like the inevitable comparisons being drawn to [the NFL]."[11] As clearly shown by this example, professional sports leagues such as the NFL, MLB, and NBA are particularly concerned with the images and narratives associated with their respective brands. Unquestionably, leagues view sports video games as an additional rhetorical resource through which they can express "who they are," which often involves political decisions over how they will be represented in their gaming counterparts. In this process, sports video games construct a particular audience, the *fan-consumer,* who will then be called upon to perform their fandom in alignment with these representations. Not only are video game advertisements positioned to induce consumption, but they also reinforce the notion that being a good fan is almost the same as being a good consumer.

The first video game advertisement is traced to Scott Adams's *Adventureland* (1978), which featured a logo of the programmer's upcoming release, *Pirate Adventure*. Early advertisements were not integrated into gameplay, but by the turn of the millennium, they had become a profitable and increasingly sophisticated business. In 2002 the South Beach Beverage Company (SoBe) became the first company to integrate its ad into the gaming content in *Tom Clancy's Splinter Cell: Double Agent,* where players could virtually purchase and consume the product with the game's main character. As "next-generation" consoles such as the Xbox 360 and the Sony Playstation 3 entered the market with wireless internet capabilities, gaming companies wielded additional flexibility in selling time- and geographically sensitive advertisements.[12] For example, Barack Obama's presidential campaign purchased ads in *Burnout Paradise* and the latest *Madden* game in the final months before the 2008 election that were visible only to online gamers who connected in so-called swing states.[13] None of this should be all that surprising considering the popularity of video games, evidenced by market research showing that Americans spent nearly $25 billion on video games in 2013, with projections indicating that the in-game advertising industry could surpass more than $7 billion by 2016.[14] Sports video games are particularly appealing; as technological advancements allowed gaming companies to re-create more of the "realism" of live sport, in-game advertisements signified a logical extension of the growing commercialism of professional and collegiate sports that constituted that realism. Furthermore, video game ads elide many of the issues associated with traditional advertising, offering "exposure to an ever-widening community [that] guarantees consumers' optimal attention."[15]

The ads themselves are but one strategy to hail fans as consumers within a sporting apparatus that spares almost all pretense that it is anything beyond a money-hungry machine with an abundance of products to sell and marketing agendas to fulfill. As a matter of fact, the ads themselves are part of a larger image that sports organizations are attempting to portray to the (potential) fans who play the video games. According to Charland, rhetoric does not address preconstituted audiences, but rather operates ideologically by compelling its addressees to subscribe to and assume a particular subjectivity that simultaneously limits and constrains their agency.[16] For Charland, audiences participate and embody the discourse that "persuades" them by consenting with the speaker's attempt at identification that exists at the foundation of all rhetorical discourses.

While video game ads could be dismissed as another form of crass commercialism that corrupts the "purity" of sport, there is evidence that gamers consent to their inclusion upon the grounds of "realism" these games allegedly replicate. A 2008 Nielsen study found that in-game advertisements yielded a "positive brand attribute association," while 82 percent of gamers surveyed indicated that video games were "just as enjoyable with ads as without."[17] Meanwhile, a recent ESPN Sports Poll showed that 75 percent of respondents believed that a sponsor's relationship to a sport was positively reinforced by the same company purchasing ads in its video game counterpart.[18] While gamers largely expressed no issues or reservations toward the opening of video games as a viable advertising platform, it is clear that the ads themselves, in general, reinforce the relationship between commercialism and sport. It is an effective strategy within a larger sports marketing agenda. This scheme wishes to position fans as consumers in order to compel purchasing behavior of the sports brand, whether it be video games, licensed merchandise, memorabilia, or sponsors' products.

However, this consent is achieved through a rhetorical sleight of hand that has long placed the burden of the growing commercialism of sports on teams' having to pay "greedy" and "narcissistic" athletes rather than the demand of high-profile companies trying to attach themselves to a dominant and visible media-driven institution. As Ron Bishop notes, "It seems that we are so busy being cynical about the players and complaining about their salaries that we leave ourselves wide open to a non-stop sales pitch."[19] Sports have become hypercommodified precisely because of the popularity they enjoy and the institutional regimes, such as team owners, media networks, and sponsors, that sense dollar signs around every unmarketed corner. Furthermore, sports video game advertisements encounter so little resistance because of the common conflation between democracy and consumption and the normalcy of living within a branded culture in which no institution, cultural activity, or leisurely escape is seemingly exempt from commercial exploitation.[20]

Thus, in-game advertisements, much like the sports video games in which they are situated, rhetorically position and constrain the audience's agency toward the prescribed subjectivity of the consumer in order to render its marketing objectives more effective. As a matter of fact, video game ads are merely one discursive resource that reinforces other narratives that operate with the same message toward the pursuit of similar objectives: namely, the reconstitution of fan identity that recognizes consumption as instrumental toward the performance of that subject position so that the institutional relationships between sports and gaming companies, media networks, and other related parties can financially maximize the return that is expected from these commercial relationships.

BRANDING THROUGH *MADDEN NFL*
FOOTBALL AND *MLB: THE SHOW*

The months leading up to the annual release of Electronic Arts' *Madden NFL* football carry with them a series of rituals that are the result of a cross-marketing spectacle without equal in the world of sports video games. There is the *Madden* cover vote coupled with a Times Square press conference broadcast live on ESPN2, which in April 2012 revealed that Detroit Lions wide receiver Calvin Johnson would grace the cover for *Madden NFL 13*.[21] Furthermore, there is the constructed mythology surrounding the *Madden* "curse," based on athletes who performed poorly or succumbed to injuries the season after appearing on the cover, that unquestionably drives part of the hype. Then there was the television show *Madden Nation*, which aired for four seasons on ESPN, featuring the top competitive *Madden* gamers competing in a head-to-head elimination tournament for a grand prize of one hundred thousand dollars. Premiering in 2005, the show was but one outcome based on a fifteen-year deal negotiated between ESPN and EA to further integrate their respective brands.[22] Finally, the annual *Madden release* is accompanied by gambling-style tournaments featuring professional gamers, some of whom are provided early copies of the game, courtesy of EA. One hard-core *Madden* professional gamer stated that "my dad would use me to hustle guys. I was 12 playing 30-year-olds. One time I lost $2,500 in a single night. My dad was disappointed."[23]

The *Madden NFL* franchise is the most popular sports title in video game history, selling more than eighty-five million copies worldwide and generating three billion dollars in revenue since it debuted on the Apple II, Commodore 64, and DOS systems in 1989.[24] Because of John Madden's status as an iconic NFL coaching legend and broadcaster, EA negotiated a deal with Madden to use his name and likeness in the game to generate brand credibility and bolster annual sales. To be sure, the hegemonic stature of *Madden* in the pantheon of sports video gaming caused advertisers to salivate in Pavlovian fashion in seeking to place their brand within the content of the game itself. As a matter of fact, the selling of advertising "space" in sports video games was not thoroughly legitimatized until EA signed a deal with Microsoft-based Massive in order to solicit ads for its top-selling sports titles such as *Madden* in 2007.[25] When Gatorade purchased ads in EA titles such as *NBA Live '09* and *NHL 10*, market research revealed that gamers who bought these titles spent an additional 24 percent on the sports drink than those who did not.[26] Not only was the beverage company benefiting from the integration of its brand identity into gaming content, but it was simultaneously reinforcing its sponsorship position with these respective leagues as the official drink of athletes and fans everywhere.

Returning to *Madden*, the prerelease hype, such as the cover press conferences and the leaking of player ratings, is the kind of staged pseudo-events intended not only to move copies of the game, but also to potentially enhance the visibility and effectiveness of the advertisements embedded within.[27] In his study exploring how Super Bowl advertisements were elevated to cultural "events," Matthew McAllister observes how companies themselves manufacture anticipation for their ads within "pre-event advertising." He notes that "advertisers hope to create advantageous 'buzz' previously reserved for star-studded motion pictures" that consequently "send the message that the event commercial is the cultural equivalent of the motion picture."[28] However, we would add that the entire manufactured "buzz" of the Super Bowl in the weeks leading up to the game creates an element of value for advertisers, as it ensures a greater supply of eyeballs that transform expensive advertising space into a worthwhile investment. Likewise, the often manufactured anticipatory eagerness surrounding the release of the latest *Madden* installment perhaps not only ensures that Electronic Arts can charge a premium for advertising space in the game, but also satisfies the will of companies who benefit from such celebratory fanfare accompanying a commercial product in which their logos can be found.

As McAllister notes, "Advertising is an overwhelmingly self-interested discourse. It is designed to sell us specific products and promote consumerism generally."[29] There is little question that the inclusion of advertisements in sports video games has the rhetorical effect of positioning audiences to assume a similar self-interested relationship with sport by linking fandom to an obligation to consume. While sports video games allow respective leagues to brand themselves within selective frames of representation, it is the limited nature of this frame that allows us to read a particular intention behind which aspects of sporting "reality" will ultimately be reflected in their video game counterparts.

In *Madden NFL 12*, one of the purposes of advertising in the game is not only to solidify existing relationships with official NFL sponsors, but to naturalize the connection between sports and consumption more generally. Upon first glance, the advertising embedded in the game appears to be minimal, as the only NFL sponsors featured prominently are the United Service Organization (or USO), Reebok, Verizon, and Gatorade, while Pepsi and Anheuser-Busch (whose combined sponsorship deals with the league top $1.8 billion) are absent.[30] While the lack of transparency in video game licensing means we can only speculate about the reasons for their exclusion, one purpose of these in-game ads is to promote the inseparability of official league sponsors and the NFL brand. In other words, *Madden NFL,* which is often heralded as the most realistic football simulation available, normalizes the association of league sponsors with the very fabric and identity of the NFL. For example, noted blogger Andy Rosenberg, who writes about in-game advertising, posted that "the point is advertising is everywhere in professional sports. The fact that it is now included in sports video games only makes the sports video gaming experience that much more authentic." Meanwhile, Elizabeth Hart, senior vice president for global media sales for EA, maintains that "part of the authenticity that is embedded in the game or that is dynamically served during connected play, includes the finer details like in-stadium and in-arena signage."[31] Thus, the ads included within sports games such as *Madden NFL 12* support these titles' claim to realism because of the authenticity allegedly demonstrated by official league sponsors appearing within the video game. It further conveys how fused their identities are with the overall NFL brand and how fans may view them as much a part of the game as the players and teams themselves.

However, to gain a more revealing and substantive view of how much advertising is included in the game, we must extend the definition of it to include not only the presence of logos and sponsors, but a general worldview that locates fan activity and identity within the very act of consumption itself. For example, the overwhelming majority of virtual spectators in *Madden* are seen sporting official team merchandise, conveying a subtle yet noticeable rhetorical suggestion that "real" fans can best identify with their favorite team by donning the same look that players wear on the field. Garry Crawford notes that "it is through consumer goods that the 'fan' can increase their knowledge, and more importantly, display their commitment through conversation and the consumer goods they own and display."[32] The NFL's official website, nfl.com, is featured visibly within the game that offers a "shop" where fans can purchase licensed merchandise in all of its colorful variations, much of it advertised in the game. As a matter of fact, cut scenes show coaches standing on the sidelines wearing some kind of logoed apparel, whether it be an official team cap, jacket, or sweatshirt. To be fair, actual NFL coaches and fans are often seen on television wearing licensed merchandise, but given the limited frames by which "reality" can be represented in the game, the inclusion of these kinds of fan and coach representations suggests a conscious message sent by both the NFL and EA that fan identity is legitimized by the consumption of the NFL brand.

While gaming companies such as EA bear some authority over which kind of content is included in the game, such as tackle physics, virtual player performance based on their actual ability, and the quality of graphics and gameplay, the NFL appears to exert discretion over how the league is represented aesthetically. Whether it is the inclusion of official league sponsors ads in the game and patriotic flair such as military flyovers before the opening kickoff, video game football becomes one avenue for the NFL to brand itself according to a selective set of images that mirror its institutional and marketing agenda. As seen in the afore-mentioned example of ESPN's *Playmakers,* deviation from a preferred narrative carries the risk of jeopardizing future contracts and licensing agreements with the league. As a matter of fact, the NFL granted EA in 2004 the exclusive rights to NFL teams, player names, and stadiums that gave them a virtual monopoly of the football gaming market.[33]

Meanwhile, the same strategies of image management are present in Sony PlayStation's *MLB: The Show 12,* where in-game advertisements serve an underlying commercial imperative behind a rhetorically crafted philanthropic veil. Like *Madden, The Show* aims to constitute a particular audience that relates to baseball through consumption, consistent with other cultural and political discourses that align democratic citizenship with the purchasing and consumption of goods and services. Many of the in-game advertisements featured in *The Show* are for nonprofit orga-nizations such as the Boys and Girls Club of America, Welcome Home Veterans, and Stand Up to Cancer, alongside for-profit companies such as Topps Trading Cards, Majestic, and State Farm Insurance Company. This juxtaposition rhetorically conflates Major League Baseball and its official for-profit sponsors with the appearance of good corporate citi-zenship and social responsibility through "cause-related marketing."[34]

According to Samantha King, "cause-related marketing" indicates a company's promise to donate a predetermined portion of their sales generated by a particular product or service to a charitable organization or cause.[35] These marketing strategies have proved exceptionally popular among consumers who gain the personal satisfaction of "intervening" on behalf of a needy cause without assuming any additional commitments beyond purchasing the product itself. As King notes, cause-related mar-keting discursively privileges neoliberal capitalism, where consumers are afforded the discretion over which causes they will support through voluntary association rather than more compulsory forms of engage-ment such as taxes. Thus, it would appear more "democratic," it makes people feel good while benefiting a needy organization, and corporations can manufacture an image of goodwill through philanthropic gestures that obscure the self-interested motivation through which they are often undertaken.

In recent years, the economic recession has led to a greater degree of public scrutiny toward corporate sponsorship of sports, some of whom were the recipients of federal bailout money.[36] In 2009 State Farm, Bank of America, and Pepsi announced that they were going to increase their charitable giving as a demonstration of goodwill and humble sensitivity in light of the current fiscal crisis. Through their sponsorship with Major League Baseball, these companies were hoping to capitalize on the game's popularity and charitable image to engender favorable public relations that would drive sales of their brand. During the 2009 Home Run Derby in St. Louis, Missouri, the vice president for research and development of Joyce Julius and Associates noted, "It is on a broadcast network, it's prime time, it draws a relatively large audience leading up to it, and has a bit of an afterlife." Many of the sponsoring companies involved in the event, such as MasterCard and General Motors, not only claimed charitable intent, but seemingly exhibited no shame in stating how their efforts were economically calculated. For example, General Motors boasted that each dollar spent on its sponsorship with Major League Baseball translated into five dollars in sales, a significant return on its initial investment. Ray Bednar, who heads global sponsorship marketing at Bank of America, said that "I can completely understand the cynicism and understand the questions about philanthropy and investment in sports." However, this cynicism had nothing to do with whether the company's charitable efforts were selflessly authentic, but rather centered on the degree to which they could bolster the bottom line. As Bednar notes, "We make money at this, and that's the most important message to get out."[37]

Similarly, in-game advertisements in *MLB: The Show* are strategically mobilized to camouflage Major League Baseball's and its associated sponsors' commercial marketing program behind a philanthropic veil of social responsibility and corporate goodwill. While charitable organizations may certainly benefit from the financial support and visibility these relationships generate, they can hardly be considered a genuinely charitable act. Meanwhile, State Farm, whose ads populate *MLB: The*

Show 12, solidifies its real-world relationship with Major League Baseball, while transforming its status as "your father's insurance company" to one deemed more hip by younger audiences through video games.[38] Like *Madden,* Major League Baseball considers video games a suitable avenue for brandishing a particular image consistent with its institutional objectives, as seen by the "philanthropic" character of the MLB being mirrored in *MLB: The Show 12.*

As in *Madden,* virtual fans in *MLB: The Show 12* are seen primarily in logoed merchandise, suggesting rhetorically that fan identity is largely predicated upon wearing officially licensed jerseys, hats, and sweatshirts. According to Bishop, teams regularly alternate uniform styles and colors between seasons with the understanding that the same jerseys worn by athletes not only are popular among fans, but also compel "real fans" to update their wardrobes in order to remain in keeping with these changing fashions. In *MLB: The Show 12,* gamers are provided the option to outfit their teams in all of these multicolor variations, from home and away to alternate and retro styles, which are then modeled by virtual fans who are seen "performing" their fan identity accordingly. As Nicholas Abercrombie and Brian Longhurst note, "Critical to what it means to be a member of an audience is the idea of *performance*" in contemporary society.[39] As performances infuse into the practices and fabric of everyday life as a consequence of the saturation of mass media texts and images, consumption becomes the primary means by which individuals demonstrate who they are to others. As *Madden* and *The Show* convey, being a "real" sports fan is no longer merely a self-designated claim, but must be put on display for the approval and sanction of others through various consumer-based acts through which contemporary identity is defined under conditions of neoliberal capitalism.

Both *Madden* and *The Show* are marketed as delivering a realistic experience of both professional football and professional baseball, which is much the same as saying that the presentation of each mirrors that of a television broadcast. From the use of real-world announcers, graphical interfaces, and camera angles, fans accept the inclusion of in-game ads to the degree that they are featured where they would normally exist during a mediated sports broadcast. In *Madden 11,* for example, both programmers and Old Spice were berated because of a "Swagger" feature sponsored by the deodorant company, where a player's performance was enhanced by excessive touchdown celebrations.[40] Had the company's logo been featured during the opening coin toss or the presentation of the starting lineups, gamers likely would have approved of their inclusion in the game. As long as ads are included in places where fans would normally be exposed to them on television, the boundaries of "reality" would not be violated. Thus, sports video games are not intended to create ad "space" where it would not normally be found, but rather should place it where it already exists.

As mentioned previously, these games validate very selective frames of representation in the terms defined by the NFL and the MLB through the various licensing agreements negotiated with companies such as EA and Sony. However, this does not suggest that the companies themselves do not maintain a degree of discretion over which content will be included, nor does it mean that they do not profit from the sale of in-game ads to league sponsors. As a matter of fact, the revenues generated from the sale of virtual ads in video games lower programming and developing expenses that reduce the cost of the product for consumers. The average fifty-dollar video game without advertising revenue yields approximately three to five dollars in profit for each unit that is sold.[41] Yet given the exclusivity of gaming licenses in which one company gains the right to use actual MLB and NFL logos, players, statistics, and other likenesses, leagues unquestionably exert much influence over how they will be represented in video games.

SPORTS, VIDEO GAMES, AND THE FAN-CONSUMER

As Benjamin Barber notes, contemporary capitalism has transformed democracy from a system that promotes self-government, political empowerment, and the common welfare to a victim of market coercion where personal liberty is associated with consumer choice. "The modern tyrant hopes to impede our aims, divert our purposes, and reformulate our goals. He is not the democratic majority or the public good, he is the enforcer of consumer capitalism's need to sell."[42] In line with Barber's vision in describing modern American society, good citizens are good consumers, adults are reduced to children to stimulate often compulsive and unreflexive behavior, and "finding oneself" is only one shopping mall trip away.

Professional and collegiate sports have blossomed into a billion-dollar industry, fueled by a business-savvy "sports-media complex" driven by an insatiable appetite to sell products, from licensed merchandise, cable packages, and sponsors' products to video games and memorabilia.[43] Thus, leagues must construct a particular relationship to sports, where fandom is less of an activity and more of a prepackaged identity not only to financially maximize these revenue streams, but also to guarantee an ever-greater monetary return for all the sponsors and other companies and industries associated with them. They must create a particular kind of audience who will not only assume the preferred consumer subjectivity but identify primarily with the NFL, MLB, and NBA through these commercialized frames of engagement. Discursive renderings of what constitutes being a "real fan" must, as Maurice suggests, interpellate individuals in following a particular script that rhetorically directs their behavior toward a particular and desired end.[44] The *fan-consumer* encapsulates this dual identity, where being a sports fan brings with it various obligations to consume in order to "correctly" perform that identity.

Unquestionably, changes in the domestic and global economy through globalization, the spread of deregulated neoliberal capitalism, and the ascendance of the corporatized state have consequently altered not only how the American public conceptualizes democracy, but its own identity as well. In other words, subjects must be encouraged to view the world through the same lens that dominant economic actors choose: to see its relationship to the world fundamentally in consumer terms. As David Harvey describes, the purpose of government within neoliberal capitalism is to establish and maintain a business-friendly climate, even in instances where the public's interest and well-being are clearly being violated. "Neoliberalism . . . proposes that human well-being can best be advanced by liberating individual entrepreneurial freedoms and skills within an institutional framework characterized by strong property rights, free markets, and free trade."[45]

While sports are often celebrated for offering fans an escape from the pressures of political and everyday life, their control by dominant interests nonetheless ensures that these ideologies will be represented and signified.[46] Thus, neoliberal thinking undergirds much of the marketing philosophy of sports leagues in reimagining the role of the fan from supporter to fan-consumer, much like contemporary democracy under similar conditions repositions citizens in similar ways. In-game advertisements are the rhetorical symptoms of a new *ethos;* not only are sponsors represented as inseparable from the NFL and MLB brands, but they encourage a relationship to sport that is dependent on consumerist engagement.

CONCLUSION

Our purposes here were to illustrate how sports video games such as *Madden NFL 12* and *MLB: The Show 12* operate as rhetorical artifacts for the MLB and the NFL to represent "who they are" as brands through the "real" sporting experiences these games allege to deliver. Because the spectrum of reality in sports video games is much more limited due to programming constraints compared to live sport, how reality is modeled in these titles involves selective frames of representation consistent with the leagues' institutional objectives. While in-game advertisements of official league sponsors may be viewed as an extension of this reality because of their visible relationship to sports, they simultaneously reinforce the notion that State Farm and Verizon are just as much a part of the MLB and the NFL brands as the players and teams themselves. The fact that in-game advertisements further "authenticate" the realism of these games is a visible example of how closely aligned these sponsorship relationships are with the identity of these leagues.

Furthermore, both in-game advertisements and the manner in which fans are represented in the game operate discursively with other sports narratives in the construction of an audience of fan-consumers. Given the growing commercialism of sports and the burgeoning number of companies who pay to be associated with them, the *fan-consumer* signifies a preferred subjectivity that views its relationship to sports in commercial terms. With identity in contemporary society commonly seen as a consumer-based "performance," fans may not view this compulsion as anything out of the ordinary.[47] Obviously, not everyone who plays a sports game like *Madden* is a fan of professional football; the focus here was on sports fans who play these games because they provide another means to interact with the MLB and the NFL. Finally, the construction of the fan-consumer does not exist in a cultural and political vacuum, but represents an extension of neoliberal-based thinking finding its expression in a dominant social institution such as professional sports. In-game advertisements are but one example where the commodification of fan identity is clearly visible and articulated.

NOTES

1. Sam Mamudi, "Study Shows Sports TV Success," *Marketwatch: Wall Street Journal,* January 24, 2012, http://articles .marketwatch.com/2012-01-24/general /30729411_1_sports-study-espnu-regional -sports; Stephen Wilson, "NBC Gets 2010, 2012 Olympics for $2.2 Billion," *Pittsburgh Post-Gazette,* June 7, 2003, http://old.post -gazette.com/tv/20030607tvrights0607p4 .asp; Brad Wolverton, "NCAA Agrees to $10.8 Billion Deal to Broadcast Its Men's Basketball Tournament," *Chronicle of Higher Education,* April 22, 2010, http:// chronicle.com/article/NCAA-Signs -108-Billion-De/65219/; Sam Mamudi, "Conference Realignment Decades in the Making: Megaconferences Mean Money and Exposure, but History Is Odd Man Out," *Marketwatch: Wall Street Journal,* September 22, 2011, http://www.market watch.com/story/college-realignments -decades-in-the-making-2011-09-21; Dennis Dodd, "College Football's Playoff Will Be Huge, and Everyone Wants a Piece of It," *CBSSports.com,* July 3, 2012, http:// www.cbssports.com/collegefootball /story/19482340/college-footballs-playoff -will-be-huge-and-everyone-wants-a -piece-of-it.

2. Linnea Kirgan, "NFL Tackles Tough Challenge of Women's Apparel," *Bizmology,* February 1, 2011, http://

bizmology.hoovers.com/2011/02/01/nfl -tackles-tough-challenge-of-womens -apparel; Jan Norman, "Business Goes to Bat for Baseball," *Orange County (CA) Register,* April 5, 2012, http://www.ocregister .com/articles/licensed-347956-products -major.html.

3. Lizabeth Cohen, *A Consumers' Republic: The Politics of Mass Consumption in Postwar America* (New York: Alfred A. Knopf, 2003), 101, 396.

4. Greg Dickinson, "Selling Democracy: Consumer Culture and Citizenship in the Wake of September 11," *Southern Communication Journal* 70 (2005): 272.

5. Robert Asen, "A Discourse Theory of Citizenship," *Quarterly Journal of Speech* 90 (May 2004): 191, 195; Richard C. Crepeau, *Baseball: America's Diamond Mind, 1919–1941* (Orlando: University Presses of Florida, 1980), 25.

6. For example, see D. Stanley Eitzen, *Fair and Foul: Beyond the Myths and Paradoxes of Sport,* 4th ed. (Lanham, MD: Rowman and Littlefield, 2009); Howard L. Nixon, *Sport and the American Dream* (New York: Leisure Press, 1984); Steven A. Riess, "Sport and the American Dream: A Review Essay," *Journal of Social History* 14 (Winter 1980): 295–303.

7. This is not a naive grasp at nostalgia for a more "pure" era when commercial

interests did not affect sport. We understand full well that no such era really existed. However, it is fair to say that especially in the so-called information (or digital) age, the synergies between sport and commercialism have accelerated at an alarming pace.

8. Lawrence A. Wenner, "Sport, Communication, and the Culture of Consumption: On Language and Identity," *American Behavioral Scientist* 53 (2010): 1572.

9. Maurice Charland, "Constitutive Rhetoric: The Case of the Peuple Québéçois," *Quarterly Journal of Speech* 73 (1987): 134.

10. Rudy Martzke, "Blowing Whistle on 'Playmakers' ESPN's Best Call," *USA Today*, February 5, 2004, http://www .usatoday.com/sports/columnist/martzke /2004-02-05-martzke_x.htm, para. 10, 8.

11. Quoted in Larry Stewart, "'Playmakers' Is Sacked by ESPN," *Los Angeles Times*, February 5, 2004, http://articles .latimes.com/2004/feb/05/sports/sp -espn2005.

12. "Entertainment Software Association," 2011, http://www.theesa.com/facts /index.asp.

13. Ahmed Ajaz, "Changing the Rules of the Game," *Campaign*, May 15, 1999, http://web.lexis-nexis.com/universe; "Entertainment Software Association"; Matthew Ingram, "Obama Targets Gamers with Ads," *Globe and Mail*, October 15, 2008, B3, http://web.lexis-nexis.com /universe.

14. Dean Takahashi, "Video Game Retail Sales Slipped in 2013 as Gamers Prepped for New Consoles," *Venture Beat*, January 16, 2014, http://venturebeat .com/2014/01/16/2013-was-a-transition -year-for-video-game-industry-retail -sales/; Paul Tassi, "Analyst Says That Video Game Advertising Will Double by 2016," *Forbes.com*, September 14, 2011,

http://www.forbes.com/sites/insert-coin/2011/09/14/analyst-says-video-game -advertising-will-double-by-2016/.

15. Amanda Andrews, "Getting In on the Game," *Allbusiness.com*, http://www .allbusiness.com/government/elections -politics-campaigns-elections/12716438-1 .html.

16. Charland, "Constitutive Rhetoric."

17. Alice O'Conner, "Study: In-Game Advertising Boosts Gatorade Sales," *Shacknews*, September 14, 2010, http:// www.shacknews.com/article/65560 /study-in-game-advertising-boosts.

18. "Study Shows In-Game Advertising Maximizes Marketing Dollars in Sports Category," *PR Newswire*, http://www .prnewswire.com/news-releases/study -shows-in-game-advertising-maximizes -marketing-dollars-in-sports-category -84019197.html.

19. Ron Bishop, "Stealing the Signs: A Semiotic Analysis of the Changing Nature of Professional Sports Logos," *Social Semiotics* 11 (2001): 30.

20. Greg Dickinson, "Selling Democracy: Consumer Culture and Citizenship in the Wake of September 11," *Southern Communication Journal* 70 (2005): 271–284; Naomi Klein, *No Logo* (New York: Picador Press, 2000).

21. Gregg Rosenthal, "Calvin Johnson Wins *Madden NFL 13* Cover Vote," *NFL .com*, April 26, 2012, http://www.nfl.com /news/story/09000d5d82898829/article /calvin-johnson-wins-madden-nfl-13 -cover-vote.

22. Thomas Oates, "New Media and the Repackaging of NFL Fandom," *Sociology of Sport Journal* 26 (2009): 31–49.

23. Quoted in Patrick Hubry, "The Full-Throttle World of *Madden*," *ESPN Page 2*, August 6, 2010, http://sports.espn.go.com /espn/page2/story?page=hruby/100806 _madden.

24. Barbara Lippert, "*Madden NFL 11:* The Ad Game," *Adweek,* July 11, 2010, http://web.lexis-nexis.com/universe.

25. Beth Bulik-Snyder, "In-Game Ads Win Cachet through Deal with EA," *Advertising Age,* July 30, 2007, http://web.lexis-nexis.com/universe.

26. O'Conner, "Study: In-Game Advertising Boosts Gatorade Sales."

27. Daniel Boorstin coined the term *pseudo-event* to refer to carefully crafted public relations ploys that were created precisely to garner media attention and the ascendance of illusion in our cultural imaginary. See Daniel Boorstin, *The Image: A Guide to Pseudo-events in America* (New York: Harper Colophon, 1961).

28. Matthew P. McAllister, "Super Bowl Advertising as Commercial Celebration," *Communication Review* 3 (1999): 411.

29. Ibid., 425.

30. Because we are examining *Madden NFL 12,* which was released in August 2011, Reebok's sponsorship deal was still in effect. Beginning with the 2012 season, Nike manufactured the official on-field jerseys for all teams as a result of a $1.1 billion apparel sponsorship. For more on the terms and conditions of many official sponsors of the NFL, see "Inside the NFL's $93.3 Billion Money Machine," *CNN Money.com,* http://money.cnn.com/galleries/2011/news/1103/gallery.nfl_total_value.fortune/4.html.

31. Andy Rosenberg, "In-Game Advertising in Sports Video Games . . . Why It Works," *Andythegiant.com,* March 12, 2009, http://andyrosenberg.wordpress.com/2009/03/12/in-game-advertising-in-sports-video-games-why-it-works; "Study Shows In-Game Advertising Maximizes Marketing Dollars."

32. Garry Crawford, *Consuming Sport: Fans, Sport, and Culture* (London and New York: Routledge, 2004), 83.

33. Tim Surrette and Curt Feldman, "Big Deal: EA and NFL Link Exclusive Licensing Agreement," *Game Spot,* December 13, 2004, http://www.gamespot.com/news/6114977/big-deal-ea-and-nfl-ink-exclusive-licensing-agreement.

34. Samantha King, "Pink Ribbons, Inc.: Breast Cancer Activism and the Politics of Philanthropy," *International Journal of Qualitative Studies in Education* 17 (2004): 473–492.

35. Ibid.

36. A prime example is the fallout that ensued when Citigroup decided to honor its twenty-year naming-rights deal for the New York Mets' new ballpark despite receiving forty-five billion dollars in bailout funds by the federal government. For more information, see Greg B. Smith and Larry McShane, "Mets Deny Report That Citigroup May Bail Out of Stadium Naming Deal," *NYDailyNews.com,* February 3, 2009, http://articles.nydailynews.com/2009-02-03/news/17916480_1_citigroup-citi-field-naming.

37. Ken Belson, "In Tough Times, Charity Sells for Sponsors," *International Herald Tribune,* July 15, 2009, http://web.lexis-nexis.com/universe, para. 12.

38. Mya Frazier, "State Farm Looks to Update Image – but Agent Spokesman Is Here to Stay; Q&A: No. 1 Player Thinks Web, Not Cavemen, Is Best Way to Up Youth Appeal," *Advertising Age,* July 9, 2007, http://adage.com/article/print-edition/state-farm-update-image-agent-spokesman-stay/119040/, para. 2.

39. Bishop, "Stealing the Signs"; Nicholas Abercrombie and Brian Longhurst, *Audiences* (London: Sage, 1998).

40. Lippert, "*Madden NFL:* The Ad Game."

41. Snyder-Bulik, "In-Game Ads Win Cachet through Deal with EA."

42. Benjamin Barber, *Consumed: How Markets Corrupt Children, Infantilize Adults, and Swallow Citizens Whole* (New York: W. W. Norton, 2007), 125.

43. For more on the "sports/media complex," see Sut Jhally, "Cultural Studies and the Sports/Media Complex," in *The Spectacle of Accumulation: Essays in Culture, Media, and Politics,* edited by Sut Jhally (New York: Peter Lang, 2006), 70–93.

44. Charland, "Constitutive Rhetoric."

45. David Harvey, *A Brief History of Neoliberalism* (Oxford: Oxford University Press, 2007), 2.

46. George Sage, *Power and Ideology in American Sport,* 2nd ed. (Champaign, IL: Human Kinetics, 1998).

47. Abercrombie and Longhurst, *Audiences.*

EIGHT

Exploiting Nationalism and Banal Cosmopolitanism: EA's FIFA World Cup 2010

Andrew Baerg

SPORT AND ITS REPRESENTATION IN MEDIA HAVE LONG BEEN A site for the communication and perpetuation of national identity. International mediated sporting events such as the Olympics and World Cup have tended to become sites allowing for the expression of myths about collective, national identities. As such, it might be expected that this tight relationship between sport and the nation-state would continue in the comparatively new medium of the sports video game, especially one representing a competition between nations.

This chapter addresses this argument by performing a textual analysis of Electronic Arts' soccer video game *2010 FIFA World Cup South Africa* (hereafter FIFA WC10) in order to learn how it positions its users. By working through and applying cosmopolitan theory and then applying this theory to the text, the chapter argues that FIFA WC10 departs from a traditionally national orientation to the mediation of world soccer toward a cosmopolitan mediation of the sport. As such, rather than position players as national subjects, FIFA WC10's various gameplay options position its users as global, cosmopolitan subjects.

SPORT, NATIONAL IDENTITY, AND COSMOPOLITANISM

Over the past two decades, sport sociologists and sports media scholars have followed the theoretical work on the nation as imagined community with invented traditions to study specific examples of the relationship between sport, media, and national identity.[1] More recent work has addressed the issue of American national identity through media coverage of the Ryder Cup golf tournament and the rhetorical construction of post-9/11 American national identity through Disney's *Miracle,* a cinematic treatment of the 1980 men's Olympic hockey victory over the Soviet Union.[2]

Sport has been understood to be particularly useful for the construction of collective identities such as national identity. Boyle and Haynes argue that sport's prominent use of symbols, focus on competition, and uniting of fan groups lend itself to an emphasis on national traits and, by extension, national identity. Sports media tend to affirm this identity by mediating between broader ideological discourses attached to national identity and the meaning of sporting events. Sports media typically articulate national team performances to a nation's perception of itself or cultural developments within that nation or both. They attach these performances to national myths as a way to express and affirm a sense of collective identity, myths that may invoke celebrations of national pride or criticism over apparent failures of national character. Even as the sports media may not be at the core of identity formation, and identities themselves exist in a state of fluidity and dynamism, Boyle and Haynes argue that the mediation of international sport has the potential to play a vital role within the broader context of collective identity formations. Mediated sport has the capacity to "reproduce, reinforce and even normalize attitudes and values which exist in other spheres of political or cultural life."[3]

To this point, the study of how these attitudes and values are reproduced and reinforced in new media like sports video games is only beginning. Baerg has addressed how sports video games position users to adopt a hegemonic masculinity, has spoken of the consequences of the games' player attribute rating systems, and discussed the sports game's relationship to history. Others such as Cree Plymire and Crawford have studied the implications of experiencing sport through the video game in the context of comparisons to other media.[4] In spite of these efforts, scholarship connecting the sports video game to media studies and sport sociology remains sparse.

One of the ways these connections might be developed concerns an investigation into the link between the sports video game and its relationship to globalizing processes. Sport, and its broader mediation, has been important for global studies given that it has energized and served as an unofficial measure of transnational change.[5] Globalization, especially the globalization that occurs through globally mediated sporting events such as the Olympics and World Cup, would appear to usher in a concurrent process of cosmopolitanization and its accompanying perspective, cosmopolitanism. Tomlinson and Young have noted how sport spectacles such as the Olympics and the World Cup can serve as important sites for engaging questions concerning the potential for cosmopolitanism that allows for competition, respect, and mutual understanding between diverse cultures. The mediation of these events via new media may open up opportunities for discussing the nature of cosmopolitanism and how it is expressed.[6]

Even as it has been deemed an epistemological necessity for twenty-first-century living,[7] cosmopolitanism has been a contested term in social theory for both its indeterminacy and its inescapability. Cosmopolitanism appears difficult to define while simultaneously serving as an increasingly prevalent part of contemporary experience.[8] This lack of clarity is furthered by cosmopolitanism's uneven development.[9] Clifford goes so far as to say that the term *cosmopolitanism* may not even speak to a coherent idea or set of experiences. In a similar vein, Pollock et al. have a difficult time ascertaining what constitutes cosmopolitanism and cosmopolitan practice. They assert that what has traditionally been understood to be cosmopolitanism has often been a product of the place from which one begins, especially with respect to the Western philosophical tradition and European intellectual history. Neatly defining cosmopolitanism, they say, "definitely is an uncosmopolitan thing to do."[10]

In spite of this uncertainty over definitions, other scholars have made a concerted effort to provide solidity to the concept of cosmopolitanism. Harvey cites Nussbaum's definition of cosmopolitanism as "a set of loyalties to humanity as a whole, to be inculcated through a distinctive educational program emphasizing commonalities and responsibilities of global citizenship." Similarly, Robbins argues that cosmopolitanism has traditionally been "understood as a fundamental devotion to the interests of humanity as a whole" and something that claims universality as a consequence of "its independence, its detachment from bonds, commitments, and affiliations."[11] If cosmopolitanism is defined around a broad interest in humanity and over and against attachments, bonds, commitments, and affiliations, then it would seem to be difficult to connect cosmopolitanism to nationalism and cosmopolitan identity to national identity. Yet many scholars have taken up this relationship to examine cosmopolitanism's relation to the nation and national identity.

Some see the ongoing importance of the nation within a cosmopolitan frame. Appiah argues for what he calls a "rooted liberal cosmopolitanism." Appiah speaks of the rooted cosmopolitan as one who is "attached to a home of his or her own, with its own cultural particularities, but taking pleasure from the presence of other, different places that are home to other, different people." This rooted liberal cosmopolitanism acknowledges difference by understanding that there is no need for a global culture of homogeneity. Given this acknowledgment of difference, the liberal cosmopolitan advocates for the freedom to create the self by drawing on a variety of social sites for the discursive and normative resources needed to create identity. The nation remains one of these important resource-providing sites. In contrast to Appiah's cosmopolitanism that remains tied to a cultural space of a national home, Ree asserts a cosmopolitanism as that which moves beyond nationality and its accompanying broader political structure of internationality. For him, cosmopolitanism allows people to think of themselves beyond a reference to the nation. Similarly, Wilson allows for a cosmopolitan perspective that is free of "particularized prejudices, fixed ties, and narrow local/national boundaries." Even in Clifford's reluctance to embrace *cosmopolitan* as a useful term, he argues that it still provides a way to think about an undercutting of ethnic absolutisms that often flow in and through nation-state, tribe, or minority.[12]

Other scholars perceive room for both national and global attachments within cosmopolitanism. Cheah locates the philosophical roots of a Western understanding of cosmopolitanism in Kant, noting that Kant's cosmopolitanism predates the contemporary nation-state. Cheah maintains that Kant's cosmopolitanism actively opposed the dynastic political system rather than the national system. As such, nationalism can still exist within Kant's cosmopolitan framework. This coexistence may not necessarily always be congenial, as popular forms of cosmopolitan consumption can oppose an oppressive national politics and the hegemony of discourses surrounding the nation-state. Cheah sees these processes fostering "popular cosmopolitanisms understood as pluralized forms of popular global political consciousness comparable to the national imagining of political community." In the same way that people once thought of themselves as part of an imagined national community, they now have the ability to imagine themselves as part of a larger and more diverse global community. Cheah is conscious about the degree to which this cosmopolitanism expresses itself experientially and institutionally in the contemporary context, but does acknowledge its possibility.[13]

This experience of cosmopolitanism has been theorized through the lens of a diversity of attachment. Robbins argues that cosmopolitan identity construction works via "a reality of (re)attachment, multiple attachment, or attachment at a distance." Cosmopolitanism recognizes and affirms the potential for connections to identities associated with one's local, national, and global communities. Given the range of attachments possible, a spatial proximity to others is not as important for identity formation within cosmopolitanism. Clifford agrees with Robbins's emphasis on the diversity of attachment. Within a cosmopolitan frame, he affirms that identity revolves around "displacement and relocation, the experience of sustaining and mediating complex affiliations, multiple attachments."[14]

Like Robbins, Anderson also emphasizes the notion of distance, but from a different perspective. Whereas Robbins argues for these various forms of attachment at a distance in the direction of the other, Anderson argues for a cosmopolitanism that emphasizes a "reflective distance from one's cultural affiliations" that thereby enables an understanding of the other. She suggests that in the twentieth century, this distancing occurred from nation, race, and ethnicity. For her, engaging in this kind of reflexivity may mean a distancing that results in either a lack of attachment to any specific culture or an acceptance of a variety of different cultures or both. As such, this form of cosmopolitan distancing forges a dialectic between exclusionary cosmopolitanism, which does not link itself to any specific culture, and an inclusionary cosmopolitanism, which accepts all kinds of cultures.[15] As a consequence, the exclusionary-inclusionary cosmopolitanism allows for both a global cosmopolitanism and a multiplicity of local cosmopolitanisms to operate simultaneously.

Beck also argues for an understanding of cosmopolitanism within the framework of the global and the national. For him the cosmopolitan is one who is involved in imagined dialogue with other cultures and rationalities such that this dialogue occurs with an internalized other. This dialogue necessitates "the negotiation of contradictory cultural experiences into the centre of activities: in the political, the economic, the scientific and the social." This dialogue suggests an elision of borders with the implication that borders no longer predetermine but exist as fluid and chosen. He suggests that globalization is a process by which the global and the local are mutually implicating. Cosmopolitanism and the process of becoming more cosmopolitan, that is, cosmopolitanization, are oriented around "*internal* globalization, globalization *from within* the national societies." As this cosmopolitanization occurs, consciousness and identity begin to take on a different shape as the global becomes imbricated in local experience. Beck is careful to assert that cosmopolitanization does not automatically mean that we will all be cosmopolitans, but this process does entail that vital questions surrounding things such as ways of life and identities can no longer be confined in the national and the local. Cosmopolitanism alters our relationship to space, time, and identity.[16]

Beck goes on to pick up on Billig's notion of banal nationalism, but adapts it to discuss what he terms "banal cosmopolitanism." He defines banal cosmopolitanism as a process by "which everyday nationalism is circumvented and undermined and we experience ourselves integrated into global processes and phenomena." Elsewhere, Beck explains the everyday nature of banal cosmopolitanism as something latent, unconscious, and passive. Banal cosmopolitanism sits behind the primary cultural signifiers that speak to national identity and consciousness. This banal cosmopolitanism is experienced in everything from youth culture to food to content in the mediascape. Banal cosmopolitanism is increasingly commodified as people buy and sell goods reflecting and reinforcing a cosmopolitan perspective. Beck argues that people do not necessarily conspicuously demonstrate "symbols of banal cosmopolitanism, but intentionally or unintentionally they do show their colours, as it were, in a cosmopolitan way."[17] These expressions of cosmopolitan become increasingly commonplace in the context of a global media and the global circulation of goods.

Studies addressing this banal cosmopolitanism have been scant. Kuipers and de Kloet's reception study of *The Lord of the Rings* film represents one of the few applications of Beck's theory.[18] Their work revisited a previous large-scale cross-national survey on the movie's fans and their response to the movie. Because of the similar ways in which diverse nations of audiences interpreted the film, Kuipers and de Kloet concluded that this audience shared more of a global, homogenous understanding of the film. In this understanding, they find evidence of a banal cosmopolitanism, one that is, in part, constructed by the Hollywood movie industry.

This chapter represents another study on banal cosmopolitanism, but focuses on the new media text of a sports video game. It builds on the scholarship above by arguing that EA's soccer simulation of the 2010 FIFA World Cup, *FIFA WC10,* exploits the tension between positioning users to consider themselves as members of an inclusionary and exclusionary cosmopolitan global community. The ensuing sections apply both Beck's notion of banal cosmopolitanism and Anderson's ideas on inclusionary and exclusionary cosmopolitanism to see *FIFA WC10* situating users within a cosmopolitan framework. Even as it is grounded in simulating a nation-based competition, *FIFA WC10* exists as an example of banal cosmopolitanism as a global media commodity circumventing and potentially undermining the very national orientation of the event it simulates. It expresses this banal cosmopolitanism by positioning users to adopt both an inclusionary and an exclusionary cosmopolitanism. By examining various aspects of its representation and gameplay options, the chapter explains how this positioning occurs and discusses some of the potential implications of this positioning for the user.

BANAL COSMOPOLITANISM IN *FIFA WC10*

To build from the work above, the rest of the chapter provides a textual analysis of EA's soccer simulation *FIFA WC10. FIFA WC10* moves between positioning the player between two forms of banal cosmopolitanism, an inclusionary cosmopolitanism that enables attachment from a distance through an acceptance of various kinds of national cultures and an exclusionary cosmopolitanism that does not attach itself to a specific national identity.

Inclusionary Banal Cosmopolitanism

Throughout various aspects of the FIFA WC10 experience, users are constantly invited and incited to select nations with which to align themselves. By allowing for this process to occur, the game positions users to identify with an inclusionary banal cosmopolitanism. Every opportunity to select a nation becomes another chance to demonstrate acceptance of another culture. This invitation occurs from the very first time the user places the disc in the console's tray and loads up the game and continues every time the user plays the game thereafter. With each loading of the game, users see an opening screen where they are asked to select their preferred language of choice: English, French, or Spanish. This choice appears every time the game is played such that FIFA WC10 encourages users to experiment with a language other than the default language assigned by region. The fact that users cannot, as in many other games, make their language choice once, never having to choose again, positions them to experiment with the different language options.

The first time users play FIFA WC10, subscribers to Xbox Live, Microsoft's online player matching system, are connected to the service. Upon this connection, gamers are asked to select a nation. This nation will become the one they represent in a global contest for online soccer supremacy. However, in this step, users are not required to select their geographical location. They are free to choose whatever FIFA-recognized nation they wish. After this choice has been made, every action the user performs in the game from this point forward makes a contribution to that user's selected nation's ranking. If a user selects the Netherlands, every game the user plays generates points that go toward the Netherlands' overall ranking on Xbox Live. Whether a user plays the game strictly as the Netherlands or with a variety of teams, each of that user's actions contributes to the overall score of the nation that has been initially selected. The same rules apply to individual actions as well. If a user plays online with nineteen other users around the world, that user's points will still be allocated to the Netherlands, irrespective of whether the user's team was the Netherlands. However, users are never tied to continuing to accrue points for their initially chosen nation. They are free to switch between nations at will by altering their game profile. A national affinity subsequently becomes an ongoing choice users can make at any point in the game experience.

The ability to be aligned with different nations continues into gameplay. Gameplay is oriented around three modes: the "2010 FIFA World Cup," "Captain Your Country," and "The Story of Qualifying." Each of these three modes enables the user to select an affiliation with a chosen nation and play as that nation over the duration of the mode. The "2010 FIFA World Cup" mode allows the user to start from the continentally based qualification stages, play through the qualifying games to gain entry to the World Cup, and then compete for the World Cup itself. The qualification stages range from six games in the Oceania region to eighteen games in South America. The second major game mode is called "Captain Your Country." In this mode, the user creates an individual avatar, assigns that avatar to any national team, and controls only that avatar as the mode unfolds. The user is then tasked with making the selected nation's first team, contributing to the team's success through the World Cup qualifying competition, and ultimately helping the country in the World Cup itself. The goal of this mode is to do well enough to be given the captain's armband and win the World Cup for the user's chosen nation. Finally, "The Story of Qualifying" mode presents the player with various historically based qualification scenarios that ask the user to equal or better a given team's performance in that match. One notorious example presents players with the opportunity to lead Ireland to victory after France tied a crucial qualifying game with help from a blatant, but uncalled, Thierry Henry handball.

No matter the mode, users are free to select any FIFA-recognized nation as their chosen squad. Whether it be Uruguay, Uzbekistan, or the United States, numerous opportunities exist for users to work through different experiences of making a national team in Captain Your Country or helping a national team on their way through the World Cup qualifying process. Due to the relatively short nature of these qualifying campaigns, users are positioned to replay them with teams from different qualifying groups and different continents. Some users may appreciate the challenge of qualifying with a middling nation in the European group, while others may choose a comparatively easier task such as qualifying with Mexico or the United States in the North and Central American group. These different roads to the World Cup and the game's diversity of represented nations position the user to accept various cultures in keeping with inclusionary banal cosmopolitanism.

The Xbox Live version of the game furthers this aspect of the user's positioning within inclusionary banal cosmopolitan by linking various achievements to an identification with those who are not part of one's home country. Achievement points are awarded for qualifying for the World Cup with a country from every continent. Achievement points are also awarded for merely playing the first stage of the World Cup online with various levels of teams. As a consequence, users receive points for playing as teams with comparatively poor ratings all the way up to playing as the best teams in the game. Considerably more points are awarded if a user is able to win an online World Cup with one of the world's soccer minnows. To be awarded these achievement points for success with countries other than one's own encourages a recognition, acceptance, and at least potential appreciation of other cultures.

Other elements of the game also contribute to this inclusionary banal cosmopolitanism. FIFA WC10 is rife with national symbols. While players wait for each segment of gameplay to begin, a splash screen appears with a map highlighting a given country, accompanied by a fact about that country. The countries featured on this screen will usually be those unaffiliated with the player's immediate competition, both in terms of the next match and within qualifying itself. The map provides a visual symbol of that nation as a kind of geography lesson for those potentially unfamiliar with the locations of Algeria or New Caledonia. It also provides factual information about the country with respect to its capital city, population, or World Cup history. To move into the core gameplay of the respective 2010 FIFA World Cup, Captain Your Country, and Story of Qualifying modes is to see a continuation of the pervasive presence of various symbols of the selected playing nations. These symbols include flags, national colors displayed in the crowds at the stadiums, nation-specific chants from these crowds in the language of that nation, national uniforms, and national anthems presented ahead of each match. Much like the maps appearing on the splash screen preceding each match, these visual and aural markers of nationality position users to recognize and appreciate the diversity of global soccer cultures and their accompanying national fan cultures.

In the same way that the game positions users to adopt an inclusionary banal cosmopolitanism through its consistent allowance of nation selection, so too is this form of cosmopolitanism affirmed by the game's soundtrack. The game's soundtrack represents an eclectic mix of styles and nations of origin. The artists hail from a diversity of places, including Senegal, Portugal, England, Netherlands, Canada, South Africa, Chile, and India, among others. It would seem that few users would recognize a majority of these artists or the songs on the soundtrack. However, repeated play in the game's menus has the potential to move them to investigate these artists further and begin to appreciate a different culture's musical genre.

Inclusionary banal cosmopolitanism is fostered by an ongoing recognition of alternative national possibilities and cultural symbols. By encouraging users to align with a variety of different cultures and their national teams, FIFA WC10 positions them to develop multiple attachments and belongings (and belongings at a distance) characteristic of an inclusionary cosmopolitanism.

Exclusionary Banal Cosmopolitanism

While FIFA WC10 positions users to adopt an inclusionary banal cosmopolitanism, it simultaneously positions them to adopt an exclusionary banal cosmopolitanism as well. Where inclusionary banal cosmopolitanism welcomes different perspectives and ideas, exclusionary banal cosmopolitanism denies the potential for connections to specific national cultures and forms of attachments. In FIFA WC10 exclusionary banal cosmopolitanism operates primarily through its hypergeneric forms of representation and through its quantification of player performance and skill.

First, exclusionary cosmopolitanism occurs in the amount of generic content in *FIFA WC10*. The generic nature of the experience is expressed in the game's media reports, player animations, and team playing styles. Whether a user chooses to play the regular qualifying or World Cup tournament or Captain Your Country modes, the game provides brief news reports assessing the user's team's performance. Typically, these stories will informally gauge an expected outcome versus how the user's match actually unfolded. As an example, if a user had chosen a soccer power such as Spain and her Spanish squad had only narrowly defeated the tiny nation of Luxembourg, the news story would talk about how the team had performed poorly in spite of the victory. Similar headlines and stories would appear for upsets when a user's team had exceeded expectations and for draws or wins when expectations had been met.

Although these respective stories do speak to the details of match outcomes, they do so in a generic manner. For a user to try to take Burundi, the Bahamas, or Brazil to the World Cup is to see the same types of news stories appear each time the game is played. These stories fail to provide any political or cultural context that differentiates one nation's experience from another. The generic nature of these stories becomes that much more important in light of the way that the sports media have played such a vital role in perpetuating discourses of national identity for self and other.[19]

Player animations also appear highly generic. During on-field game-play sequences, each team in the game is allocated a generalized assortment of dribbling, shooting, passing, and tackling options. Some of the more talented players on the better teams will have a wider variety of trick moves and more of an ability to successfully shoot, pass, and tackle than less talented players. However, the available actions remain relatively similar irrespective of the team chosen. Every one of the game's players dribbles, kicks, and passes the ball the same way. Additionally, each time the user takes control of a given player, a generic bar at the bottom of the screen indicates the name of that player, the player's stamina, and his national affiliation. These labels do not exist as stylized in a particular, country-specific font but are the same across each nation in the game.

The generic nature of the animations carries over into the generic playing styles of the respective national teams. Although it is entirely possible for users to edit each team's tactics individually (an arduous, tedious task), on the game's default settings every team plays with the same style. Different teams will play with different formations depending on their personnel, but without exception every national team employs a quick, pressing, short-passing style that, in reality, is used by only the most skilled and physically capable teams in the world. With Vanuatu playing the same way as Russia, Chad, and Venezuela, FIFA WC10 defaults to a supranational playing style that departs from any concrete link to specific national styles such as Brazil's "samba football" or Holland's "total football" or Spain's "tiki-taka" style. To watch a FIFA WC10 match without knowing who was playing would be to see a very standardized and, due to the nature of the event being simulated, globally similar style of soccer.

To look at the generic media reports, generic animations, and generic team playing styles is to see a simulation that denies any concrete links to specific national cultures. The game assumes this generic layer of representation can be laid over top of what have traditionally been more culturally specific expressions. These representations subsequently become emblematic of banal exclusionary cosmopolitanism.

These forms of generic representation are not the only way in which the game positions its users within a banal exclusionary cosmopolitanism. FIFA WC10 also positions its users within this frame through its player performance and attribute rating systems. Both systems effectively impose a global, quantitatively oriented surveillance upon the population of the game's soccer players. Each and every player in the game from all of FIFA's 199 real-world teams is subjected to both an individual rating for each match the player plays and a broader skill and ability rating representing an aggregate of the player's physical, technical, and mental prowess.

At the match level, players are constantly assessed during the game on a 1-to-10 scale, with 1 being the worst and 10 being the best. Points are awarded and deducted based on a given player's performance in the match. Successful passes, tackles, and shots will yield a higher match performance score, while fouls, cards, and missed passes will yield deductions. The match rating system does not discriminate in its application to the global category of the soccer player. Every single match a user plays, no matter which team the user controls or which opponent has been selected, invokes the match rating system. No distinctive, nation-specific systems of rating can be invoked by the user. Strikers from Burkina Faso are measured by the same system as defenders from Costa Rica. As such, a link to a culturally specific means of assessing individual player performance is implicitly denied in the face of what is presented as a universal means of tracking what players have and have not done.

In delegating player attributes, every player is assigned a number in a variety of skill categories from a low of 1 to a high of 100. Among the thirty-three ranked attributes are acceleration, dribbling, free-kick accuracy, positioning, shot power, and tactical awareness. As with the match rating system, this player attribute rating system exists as a global assessment grid laid over players from each nation. A variety of culturally specific forms of assessing player skill is nonexistent, as the far-reaching attribute rating system touches each player represented in the game the same way. Because the attribute rating system is applied to every player the same way, it subsequently positions users to perceive it as a supranational form of player measurement as opposed to the arguably Western form of assessment it represents.[20]

Whether it be the highly generic nature of its media reports, player animations, and team playing styles or the all-encompassing grid of quantitative rationality imposed on virtual soccer players via the match and player attribute rating systems, FIFA WC10 positions users within the parameters of an exclusionary banal cosmopolitanism that denies overt connections to nation-specific practices.

IMPLICATIONS

FIFA WC10 positions its users to simultaneously adopt both an inclusionary banal cosmopolitanism, representing an acceptance of a diversity of cultures, and an exclusionary banal cosmopolitanism, which de-emphasizes attachment to particular national cultures. What follows from the presence of these two forms of banal cosmopolitanism?

In terms of inclusionary banal cosmopolitanism, the persistent incitement to nation choosing combined with the game's eclectic soundtrack illustrate that identities are not linked to geography. Clearly, the game's designers could have created a default in which gamers would play with the nation where their console is registered. It could even have confined the number of playable teams to only those who had qualified for the 2010 World Cup instead of representing every one of the 199 teams involved in FIFA. But the game does not do this, and as a result users may develop an attachment to nations having little to do with their place of origin as a result of playing the game. If a gamer lives in the United States but would like to play as Wales in online play, she can do so. This American may be motivated to learn more about the Welsh as a consequence of trying to qualify for the World Cup with a nation she may know little about. As such, the game allows for a potential connection to others across distance even if this connection is only imagined.

At the same time, the game's affordance of national choice implicitly recognizes the possibility that some of its users may already possess diverse forms of attachment. FIFA WC10's design affirms that those who play the game may have attachments to potential places of origin or ancestral places of origin. If a user lives in France but has ties to the Ivory Coast, he can choose to compete on behalf of the African side known as "Les Elephants." If a user lives in Brazil but maintains a connection to Portugal, controlling the likes of Cristiano Ronaldo becomes a real possibility. FIFA WC10 freely enables a transnational deterritorialization to occur with this choice. Again, this expression of inclusionary banal cosmopolitanism enables imagined community with others outside of one's own immediate geographic space.

With respect to exclusionary banal cosmopolitanism, the highly generic nature of FIFA WC10's media reports, player animations, and attribute rating system has consequences for the user's perception of soccer. Although the media reports add some depth to the experience of qualifying and to the tournament itself, reading the same bland stories applied to the same situations renders their inclusion relatively meaningless. As stated above, the mediation of national soccer matches has played an important role in the formation of national identities over the past century. These generic media reports position users outside of the uniquely colorful representations of the game that have been produced in the past.

Additionally, the quantitative systems of player assessment, at the levels of both match and ability, present themselves as a global standard of measure in FIFA WC10. However, even as the game represents these numbers as a universally applicable system and thereby positions users to perceive them as universal, these ratings systems and their thoroughly statistical orientation normalize a very Western epistemology. Over the past several centuries, numbers have played an important role in the West's economic and political development.[21] Given the Western form of vision afforded by numbers in a game such as FIFA WC10, it appears that banal exclusionary cosmopolitanism has the potential to obfuscate the more culturally specific nature of certain epistemologies and represent them as something global.

Whether it be inclusionary or exclusionary banal cosmopolitanism, this discussion must acknowledge that forms of cosmopolitanism may foster resistance. Certainly, those who play FIFA WC10 may participate in a restricted tribalism in which they do nothing but play the game as their birth nation. They may devote all of their efforts to doing everything they can to see their nation at the top of the overall global rankings. They may play through the qualifying tournament and the Captain Your Country modes several times with their birth nation. However, in doing so, they miss out on the wider range of ways to play the game and some of the rewards provided to those who step outside of this narrower form of play. Users may also recognize some of the more generic aspects of the game's mediation and wonder why these elements of the experience appear so bland. In doing so, users may work against FIFA WC10's inclusionary or exclusionary banal cosmopolitanism.

Given the potential resistance to the game's positioning, this chapter subsequently encourages a direct study of users who play sports video games involving national competition. Where many international sporting competitions are still mediated through a national mass communication lens, the sports video game serves as a site where this national lens is potentially replaced by an alternative. How users respond to this difference and how it shapes their perception of their own national and potentially extranational identities represent topics for further study.

CONCLUSION

This chapter has examined cosmopolitan theory and applied this theory to EA's soccer simulation *FIFA WC10*. It has explained how the game positions users within the framework of inclusionary and exclusionary banal cosmopolitanism. With its encouragement of aligning the user with a variety of different cultures and their national teams, *FIFA WC10* positions users to develop multiple attachments and belongings characteristic of inclusionary cosmopolitanism. With its generic media representations, animation, playing styles, and forms of player assessment, the game positions users to perceive soccer as supranational and above specific attachments to culture and nation.

The types of banal cosmopolitanism discussed here may not necessarily be confined to games like *FIFA WC10* but might be extrapolated to many different contemporary game genres that enable the possibility of multiple attachments and attachment at a distance. The chapter raises questions about the tensions existing between inclusionary and exclusionary forms of cosmopolitanism and their interaction within new media. Whether the sports video game, or the video game more broadly, allows for the expression of cosmopolitanism is ultimately open to question.

NOTES

1. Benedict Anderson, *Imagined Communities: Reflections on the Origin and Spread of Nationalism* (London: Verso, 1983), 6; Eric Hobsbawm and Terrence Ranger, eds., *The Invention of Tradition* (Cambridge: Cambridge University Press, 1983). For some examples, see Neil Blain and Raymond Boyle, "Battling along the Boundaries: The Marking of Scottish Identity in Sports Journalism," in *Scottish Sport in the Making of the Nation: Ninety-Minute Patriots,* edited by Grant Jarvie and Graham Walker (New York: St. Martin's Press, 1994), 125–141; Ben Carrington, "'Football's Coming Home' but Whose Home? And Do We Want It? Nation, Football, and the Politics of Exclusion," in *Fanatics! Power, Identity, and Fandom in*

Football, edited by Adam Brown (London: Routledge, 1998), 101–123; Liz Crolley and David Hand, *Football, Europe, and the Press* (London: Frank Cass, 2002); Neil Earle, "Hockey as Canadian Popular Culture: Team Canada 1972, Television, and the Canadian Identity," in *Slippery Times: Reading the Popular in Canadian Culture,* edited by Joan Nicks and Jeannette Sloniowski (Waterloo, Ontario: Wilfred Laurier Press, 2002), 321–344; Richard Gruneau and David Whitson, *Hockey Night in Canada: Sport, Identities, and Cultural Politics* (Toronto: Garamond Press, 1993); Richard Giulianotti, "Built by the Two Varelas: The Rise and Fall of Football Culture and National Identity in Uruguay," in *Football Culture: Local Contests, Global Visions,* edited by Gerry P. T. Finn and Richard Giulianotti (London: Frank Cass, 2000), 134–154; Richard Holt, "The King over the Border: Denis Law and Scottish Football," in *Scottish Sport in the Making of the Nation,* edited by Jarvie and Walker, 58–74; Daniel S. Mason, "'Get the Puck Outta Here!': Media Transnationalism and Canadian Identity," *Journal of Sport and Social Issues* 26 (2002): 140–167; Michael Oriard, *Reading Football: How the Popular Press Created an American Spectacle* (Chapel Hill: University of North Carolina Press, 1993); and Nicola Porro and Pippo Russo, "The Production of a Media Epic: Germany vs. Italy Football Matches," in *Football Culture,* edited by Finn and Giulianotti, 155–172.

2. John Harris and Mark Lyberger, "Mediated (Re)presentations of Golf and National Identity in the United States: Some Observations on the Ryder Cup," *International Journal of Sport Communication* 1 (2008): 143–154; Michael L. Butterworth, "Do You Believe in Nationalism? American Patriotism in *Miracle,*" in *Examining Identity in Sports Media,* edited by Andrew

C. Billings and Heather L. Hundley (Los Angeles: Sage, 2010), 133–152.

3. Raymond Boyle and Richard Haynes, *Power Play: Sport, the Media, and Popular Culture* (London: Addison-Wesley, 2009), 144–148, 153.

4. Andrew Baerg, "Fight Night Round 2: Mediating the Body and Digital Boxing," *Sociology of Sport Journal* 24 (2007): 325–345; Andrew Baerg, "Classifying the Digital Athletic Body: Assessing the Implications of the Player-Attribute Rating System in Sports Video Games," *International Journal of Sport Communication* 4 (2011): 133–147; Andrew Baerg, "Digital Hoops History: *NBA 2K12* and Remediating Basketball's Past," *Communication and Sport* 1 (2013): 365–381; Darcy Cree-Plymire, "Remediating Football for the Posthuman Future: Embodiment and Subjectivity in Sports Video Games," *Sociology of Sport Journal* 26 (2009): 17–30; Garry Crawford, "'It's in the Game': Sport Fans, Film, and Digital Gaming," *Sport in Society* 11 (2008): 130–145.

5. Richard Giulianotti and Roland Robertson, "Sport and Globalization: Transnational Dimensions," *Global Networks* 7 (2007): 108.

6. Alan Tomlinson and Christopher Young, "Culture, Politics, and Spectacle in the Global Sports Event: An Introduction," in *National Identity and Global Sporting Events: Culture, Politics, and Spectacle in the Olympics and the Football World Cup,* edited by Alan Tomlinson and Christopher Young (Albany: SUNY Press, 2006), 1.

7. Ulrich Beck, "Cosmopolitical Realism: On the Distinction between Cosmopolitanism in Philosophy and the Social Sciences," *Global Networks* 4 (2004): 133.

8. Scott L. Malcomson, "The Varieties of Cosmopolitan Experience," in *Cosmopolitics: Thinking and Feeling beyond the*

Nation, edited by Pheng Cheah and Bruce Robbins (Minneapolis: University of Minnesota Press, 1998): 233–245.

9. Rob Wilson, "A New Cosmopolitanism Is in the Air: Some Dialectical Twists and Turns," in Cosmopolitics, edited by Cheah and Robbins, 353.

10. James Clifford, "Mixed Feelings," in Cosmopolitics, edited by Cheah and Robbins, 362; Sheldon Pollock et al., "Cosmopolitanisms," Public Culture 12 (2000): 577.

11. David Harvey, "Cosmopolitanism and the Banality of Geographic Evils," Public Culture 12 (2000): 530; Bruce Robbins, "Actually Existing Cosmopolitanism," in Cosmopolitics, edited by Cheah and Robbins, 1.

12. Kwame Anthony Appiah, "Cosmopolitan Patriots," in Cosmopolitics, edited by Cheah and Robbins, 91, 94, 97; Jonathan Ree, "Cosmopolitanism and the Experience of Nationality," in Cosmopolitics, edited by Cheah and Robbins, 88; Wilson, "New Cosmopolitanism," 354; Clifford, "Mixed Feelings," 365.

13. Pheng Cheah, "The Cosmopolitical – Today," in Cosmopolitics, edited by Cheah and Robbins, 22, 24, 36; Anderson, Imagined Communities, 6.

14. Robbins, "Actually Existing Cosmopolitanism," 3; Clifford, "Mixed Feelings," 369.

15. Amanda Anderson, "Cosmopolitanism, Universalism, and the Divided Legacies of Modernity," in Cosmopolitics, edited by Cheah and Robbins, 267, 268.

16. Ulrich Beck, "The Cosmopolitan Society and Its Enemies," Theory, Culture, and Society 19 (2002): 18, 17, 29–30.

17. Michael Billig, Banal Nationalism (Thousand Oaks, CA: Sage, 1995); Beck, "Cosmopolitan Society," 28; Ulrich Beck, "Cosmopolitical Realism: On the Distinction between Cosmopolitanism in Philosophy and the Social Sciences," Global Networks 4 (2004): 131–156.

18. Giselinde Kuipers and Jeroen de Kloet, "Banal Cosmopolitanism and The Lord of the Rings: The Limited Role of National Differences in Global Media Consumption," Poetics 37 (2009): 99–118.

19. See Crolley and Hand, Football, Europe, and the Press, for an extended discussion.

20. See Baerg, "Classifying the Digital Athletic Body," for a more thorough examination of the sports game's player attribute rating system and its implications.

21. See Alfred W. Crosby, The Measure of Reality: Quantification and Western Society, 1250–1600 (Cambridge: Cambridge University Press, 1997); Alain Desrosieres and Camille Naish, The Politics of Large Numbers: A History of Statistical Reasoning, translated by Camille Naish (Cambridge, MA: Harvard University Press, 1998); Ian Hacking, "Biopower and the Avalanche of Printed Numbers," Humanities in Society 5 (1982): 279–295; Nikolas Rose, Powers of Freedom: Reframing Political Thought (Cambridge: Cambridge University Press, 1999).

NINE

Ideology, It's in the Game: Selective Simulation in EA Sports' *NCAA Football*

Meredith M. Bagley and Ian Summers

ON JULY 9, 2013, THE LEADING SPORTS STORY IN TUSCALOOSA, Alabama, a college town obsessed with its university's football team, was not predictions for a third straight national championship, not news of yet another five-star recruit, nor updates on injuries and summer training sessions. Instead, inch-high headlines announced "GAME ON: EA Sports Releases *NCAA Football 14*."[1] Above the text, a color screen shot from the game featured an offensive player in the familiar crimson-and-white jersey breaking tackles on the way to a presumed touchdown. The would-be tacklers happened to be in white and maroon, the colors of Texas A&M, the only team to hand Alabama a loss in its 2012 national championship season. Though completely digital, fabricated, and based on advanced computational formulas, the video game redemption offered by the photo perfectly illustrates the power of simulation-based digital games such as EA Sports' *NCAA Football*.

This chapter assesses the "dynasty" mode of gameplay within EA Sports' *NCAA Football* franchise of games. In this mode, gamers act as head coach of a program, managing nearly all aspects of a team: recruiting high school seniors, cutting players after each season, organizing a playbook, and even calling plays during the games. Teams win or lose games and subsequently get invited to various bowl games or stay home altogether based on their record. In turn, this affects whether the gamer is retained or fired by their institution or receives offers from other schools. True to its name, dynasty gamers can take their team decades into the future if they wish. Assessing this aspect of the game, we address the thorny question of ideology and simulation. Despite its incredibly advanced simulations and mountains of statistics, *NCAA Football* is not a pure simulacrum of college sport, and thus ineligible for ideological critique. In this chapter, we argue that the careful inclusions and exclusions of "reality," authorized by the NCAA, make it a text steeped in ideology. After engaging with simulation theory and digital game scholarship, we suggest that despite Baudrillard's admonitions to the contrary, it is possible to identify third-party ideological interests within highly simulated media.

We suggest a dynamic of *selective simulation,* wherein sponsors or licensees exert influence of the aspects of "reality" simulated in their games. This selective simulation is different, we argue, than the general dynamics of modeling or simulation. The realism presented in *NCAA Football* is shaped by selectivity, and it follows a different theoretical arc than a pure simulacrum analysis. We ask whose interests are served by the level of simulation allowed by the NCAA within the game, and what the effects or implications might be for avid players of the game, especially if they are also avid fans of actual college football. The EA Sports *NCAA Football* game provides an ideal case study because it is a leader among licensed simulation games.

EA Sports' *NCAA Football* series is one of the most prominent sporting game franchises on the market today. As the sole owner of the NCAA license, it is the only football video game that can feature actual universities, stadiums, bowl games, and fight songs. Due in large part to this status, *NCAA Football* is consistently one of the best-selling games of its genre. When *NCAA Football 2012* hit the shelves in July 2011, it sold more than seven hundred thousand copies in the first two weeks alone.[2] All of this is indicative of a larger trend of the growing popularity – and subsequent profitability – in college sports. In the 2010–2011 academic year, the National Collegiate Athletic Association estimated its member institutions generated more than six billion dollars in revenue from athletics alone.[3] *NCAA Football* is the longest-running collegiate gaming franchise, issuing new editions for twenty-one consecutive years. The dynasty mode was introduced in 1997 and proved wildly popular. Within two weeks of the game's release in 2011, more than fifty thousand gamers launched dynasty franchises in the online version of the game alone, with countless more managing teams offline.[4] Estimates range from 70 to 90 percent of *NCAA Football* gamers play in dynasty mode. Finally, EA's company motto, "It's in the game," brags of total accuracy and realism. Critical assessment of the game, however, brings us to the question: what reality is allowed to be simulated?

SIMULATION, IDEOLOGY, AND DIGITAL GAMES

Simulation theory has become a powerful theoretical framework for video game scholars. Most closely associated with the work of Jean Baudrillard, thinkers both prior and contemporary to him have explored the shifts in experience, knowledge, and consciousness that have occurred as mass-produced images have come to dominate our lives.[5] Bringing these dense theoretical works into the applied study of digital games has accelerated in recent years. Many game studies point to Aarseth's work on cybernetics as an early bridge between traditional ludological studies – the study of games and play – and emerging uses of simulation theory.[6] Frasca urged game scholars to focus upon simulation in their assessments of effects and functions of games, though he defined simulation more simply. For Frasca, the simulation function of games meant that they "model a (source) system through a different system which maintains to somebody some of the behaviors of the original system."[7] A focus on behavior was significant, Frasca argued, for setting digital simulation games apart from narrative-based games with plotlines and character trajectories.

Giddings took Frasca's call several steps further, engaging more directly with simulation theory and how it could elevate game studies. Giddings considers digital games a hybrid cultural form that emphasizes features and structures in a unique way from its board game predecessors. The dynamic and procedural nature of digital games required more advanced analytical approaches, Giddings argued, since "computer simulations can be seen not only as the visual presentation of artificial realities . . . but the generation of dynamic systems and economies."[8] The generative power of simulation games is to be taken seriously, Giddings argues, with implications well beyond the technical aspects of game design.

Giddings's warnings of a "hyperreality" created by the "sheer pro-
liferation of television screens, computer networks, theme parks and
shopping centers" rely upon a more robust engagement with simulation
theory than many of his scholarly peers pursue. Indeed, Baudrillard's
notion of simulation pushes beyond the reality-representation dynamic
that we often connect with processes of modeling and simulation. That
is, a game (or anything else) that endeavors to replicate some system or
entity or experience can be called a simulation – the player of the game
gets to feel like he or she is "doing" the thing despite not actually being
in that situation. We can trade properties in the board game Monopoly,
explore castles in a video game, build and manage whole towns in *Sim
City,* or conduct global war old-style in the dice game Risk. In all these
cases, we are not actually doing these actions, but a simulated version
of them, with certain omissions, fabrications, or models in place to help
us feel as close as possible without leaving the comfort of our couch or
game room. The nonreality of these "realistic" or simulated games is
central to their existence and success. What Giddings and Baudrillard
are concerned about is when simulations begin to layer and reference
each other rather than some objective "real" world upon which accuracy
or realism can be measured.

This layering of simulations is needed, Baudrillard argues, and is
crucial to understanding the power of images in today's advanced society. For Baudrillard, we live in an era "inaugurated by a liquidation of
all referentials," meaning we are so immersed in models and simulations
that we think these fabrications *are* the real estate market, castle, city, or
global war. Further, and worse, in his view, the "artificial resurrection"
of reality within a system of signs has created "a space whose curvature
is no longer that of the real, nor that of the truth." Borrowing terms
from classical rhetoric and philosophy, Baudrillard warns of proliferat-
ing "simulacra": a world where signs (symbols, representations) can no
longer be swapped out with a reality that they were once connected to,
a world where signs now swap with more signs, a world that resembles a
soup of symbols and models that he calls the "hyperreal." Baudrillard,
citing the poet Jorge Luis Borges, provides a metaphor for his theory: a
map of the world that was once used to direct us, via symbols, to new
areas, spaces, realities, but has now become more real than the spaces it

used to represent. As Cubitt summarizes, today the map precedes the territory, and the symbols tell us what is real, rather than reality guiding the formation of symbolic artifacts like maps.[9] In perhaps simpler terms, the hyperreal is generated by models of itself.

In his most succinct presentation of the way layering simulations can create a new level of reality, Baudrillard offers that "simulation envelops the whole edifice of representation itself as a simulacrum" that proceeds through four levels:

1. The image is the reflection of a profound reality.
2. The image masks and denatures a profound reality.
3. The image masks the absence of a profound reality.
4. The image has no relation to any reality whatsoever: it is its own pure simulacrum.

In this chapter, we suggest that the success and powerful simulation capability of games such as EA Sports' *NCAA Football* can be described with this progression of simulacrum development. We want to explore what may happen if:

1. the media discourse related to college football (television, print, radio, internet news, and discussion) reflects an actual existence and operation of college football programs under the NCAA purview – the on-the-ground "profound reality" of college football;
2. the elements of "reality" allowed into EA's game by the NCAA provide a masking and denaturing version of college football;
3. the elements of "reality" that are not allowed by NCAA into the EA game mask the absence of any "real" college football;
4. the avid gamer's experience of college football, through the EA game, loses all relation to reality and becomes a full simulacrum.

Throughout these levels, we also want to ask about ideology – whose interests are at stake, and who benefits or loses if this progression of simulation unfolds?

The question of ideology is one of many vast, perplexing questions within simulation theory that challenges many of our taken-for-grated assumptions about how our daily lives unfold. Indeed, despite Baudrillard's wide influence, many scholars stop short of concluding that we exist today in fully hyperreal times. Many of these scholars hold reservations about how, or if, Baudrillard's theories of simulation can explain the workings of ideology in society. Although reviewing the extensive literature on definitions of ideology is not possible here, most scholars agree that powerful actors and institutions exert unequal influence over a variety of topics and processes in almost all societies through various types of ideologies. Expectations for gender roles with family structures, nationalistic standards for patriotism and citizenship, assumptions of class mobility and economic success, widely held notions on topics such as these are often influenced by ideologies that favor those in power, be it governments, economies, or institutions. The status of ideology matters because it helps reveal the workings of power and thus a starting point for agents seeking to alter conditions of society.

Baudrillard, however, argues that ideology (or ideological critique) becomes moot in a situation of hyperreality. In an important footnote to *Simulacra and Simulation,* Baudrillard explains that in simulation, as opposed to dialectical systems where two more points of exchange can be identified, "power is something that circulates and whose source can no longer be located, a cycle in which the positions of dominator and dominated are exchanged in an endless reversion that is also the end of power in its classical definition." Cubitt is equally sanguine is his assessment of social change possibilities: "In the world of simulation, there will be no revolution. Not only is truth debarred from us in the present; we cannot even look forward to its revelation in the future."[10] It is not that power does not exist in this simulacra world, but that it cannot be detected and thus altered or challenged. If Baudrillard is right, then scholars are limited, hamstrung from calling out power in the tradition of critical scholarship and pedagogy.

What do game scholars say about all this? Many of them challenge the totalizing, nihilistic conclusions of simulation theory by pointing out clear instances of ideological power within games. Leonard raises important objections to depictions of race and racialized bodies within digital games. He calls video games "powerful instruments of hegemony, eliciting ideological consent through a spectrum of white supremacist projects." Drawing on Omi and Winant, Leonard points to the way characters and design choices within video games create and extend "racial formations" of nonwhite bodies, including hyperathletic depictions of male athletes in sport games. In addition to the visual portrayal of black athletic bodies, Leonard is concerned about how "sport games indulge white pleasures as they affirm stereotypical visions of black bodies as physical, aggressive, violent." Chan has brought these types of concerns directly to game designers, asking ethical questions about "the complicity of these e-games in reinforcing hegemonic notions of power, privilege, and inequality."[11] Many of these critical race scholars cite African American politician Adam Clayton Powell III's 2003 comment that video games function as "high tech black face," extending a problematic American minstrel tradition.[12]

Scholars more focused on game design and theory approach the ideology question differently. Rather than insist upon realities and inequalities that can be exacerbated by symbolic forms, some game scholars propose that games function in a distinct way and thus are exempt from, or less susceptible to, the flattening effects of simulation theory. Frasca offered four levels of game design in which ideological interests can insert: the narrative level of representation (character design, settings, and so forth), the rules of a game (player manipulations and limits), the goals of a game (what to do to win), and then "meta-rules" that govern how players may alter rules within the game.[13] Connecting Leonard and Frasca, sport game design could be assessed as ideological in how non-white bodies are depicted (representation), what skills or actions they are allowed to perform (rules), or what incentives exist within the goals of the game for winning.

Bogost's concept of "procedural rhetoric" has an important connection to the question of ideology because it brings attention to the communication impact of gameplay. For Bogost, the computational aspect of digital games is their essence, quite literally, and marks them as unique from other forms of play or gaming. He calls this "procedural expression," or "the construction and interpretation of a symbolic system that governs human thought or action." The focus on processes, techniques, logics, and systems is also rhetorical, or fundamentally persuasive, Bogost claims. Baerg echoes this procedural emphasis with his notion of games as "active nonlinear texts" that are experienced differently each time by participants. Procedural rhetorics wield such influence that they take precedence, most often, over images within the game.[14] Images and visuals are important to the persuasive power of games, to be sure, but Bogost hones in on the systematic steps and procedures that characterize simulation games as their most dominant feature.

The relative power of procedures versus images is crucial for this study. We argue that procedures exert a level of control on a game that can be tracked to designers, licensed sponsors, and promoters.[15] In this way, procedures can be ideologically critiqued. Images, in our view, are far more susceptible to Baudrillard's levels of simulation and are far more difficult to challenge in the name of ideological critique. In many ways, our analysis still hews to Baudrillard's insights: models and simulations are, by definition, not exact, complete replicas of a "real" base subject. Fans of college football who play *NCAA Football* at home on their couches are not typically Division I college football athletes. Even when members of college football teams play the game, they too are in the comforts of home, not sprinting and colliding and competing on a field. This gap between model and real is the starting point for Baudrillard's concerns that symbols start to dictate reality in a system of simulacra. Bogost's procedural focus offers another path, however: he writes of the "simulation gap," or the space between rule-based representation and player subjectivity.[16] For Bogost, this is where critique and challenge can occur from the gamer; for us, we see this as an opening for ideological critique of the video game itself. In this study, we inquire into the space between rule-based representations and the *rule-maker's* subjectivity or ideological interests.

TEXT: EA SPORTS' *NCAA FOOTBALL*, DYNASTY MODE

In this study we assess ideological implications in the intense simulation of the "dynasty" mode of the EA Sports video game *NCAA Football*. When a player starts the *NCAA Football* dynasty mode, several steps must be taken before beginning the play of a simulated game. He or she either picks one of the coaches available from the game program – who are based loosely on their real-life NCAA counterparts – or creates his or her own coaching avatar. Each coach has a name, skin tone, body type, style of dress, face, age, and even alma mater. After the coach is selected, the player then can choose from any one of the more than one hundred schools that play collegiate football on the Football Bowl Subdivision level, the highest amateur echelon in the United States. Players can also customize conferences – while not allowed to change conference names or dissolve them altogether due to licensing limitations, they can alter their school membership, divisions, and the location of their respective conference title game. After this initial setup, players are then able to take over their team and begin the quest for college football glory.

The game then opens up in the preseason. Here, players set up their board to target potential high school recruits, establish their schedule for the year, determine who their starters and backups will be, and set their playbook. After the preseason, the game then progresses on a weekly basis during the regular season, where players can scout and "call" recruits, play the game against the scheduled opponent for that week, and make any in-season adjustments to the playbook or roster. Players can even peruse ESPN to see how teams are nationally ranked, who the Heisman Trophy favorite is at the moment, and read other headlines generated from in-season results. At any point, the player can elect to advance to the next stage of the game, wherein any elements the player did not engage in (such as recruiting or playing the football games) will be randomly simulated by the system.

METHOD: CRITICAL TEXTUAL ANALYSIS

To conduct our analysis, the author with high familiarity with the game first gathered observations from gameplay about aspects of the NCAA-approved simulation of NCAA football. The co-authors then played and explored the game together to identify any additional elements or details of design, gameplay, and procedure that were significant for the analysis. By comparing analysis throughout the process, biases were checked and gaps in observation remedied.

Our analysis technique more closely resembles Conway's "multi-level critical qualitative analysis." Conway, in turn, cites Krzywinska for guidance that we too followed: detailed textual analysis requires attention to a game's formal constructs, design environment, images, audio, capabilities of in-game objects, and characters. In addition, we followed Bogost's lead to pay close attention to procedural incentives: for instance, in assessing a game meant to criticize agrobusiness practices that support the fast food industry.[17] A blend of critical textual analysis and procedural awareness guided us throughout.

Several critical elements within the gameplay of *NCAA Football* demonstrate how the NCAA masks and denatures the on-the-ground functions of big-time college football: recruiting, player discipline, finances, and academics. Rather than a highly realistic simulation of the very complicated and difficult situations that NCAA members face on a regular basis, the partial inclusion and careful characterization of these elements move us closer to a hyperreal version of college sport.[18]

Recruiting is critical to building a successful program in the game, and the way recruiting operates within the game's system is reflective of the idealized simulation sanctioned by the NCAA. The gamer is allotted ten hours per week to call recruits, allowing up to a maximum one hour each to talk to an athlete. Gamers can then "pitch" the recruits based on more than a dozen criteria related to their school and program, such as academic prestige, the university's proximity to the athlete's home state, how often their team plays on national television, possible playing time for the recruit, the perceived quality of life on the school's campus, the quality of the athletic facilities, and so forth. The recruits – by random designation through the game's computations – will value some criteria more than the others. For instance, a recruit may not care at all about staying close to home, but makes academics his utmost priority. Based on the combination of the school's strengths in certain areas and how much the athlete values them, "points" are added to the recruitment meter. If the gamer can max out the athlete's recruitment criteria and offer him a scholarship, then that recruit "commits" and is forever signed with the gamer's team.

What is striking is that nowhere within this system are gamers ever allowed to deviate from NCAA rules. The gamer cannot exceed the allotted time for scouting or recruiting or offer an athlete any sort of benefits in exchange for picking a program. This is indicative of Bogost's notion of "procedural rhetoric," in that what systems allow or disallow players to do can reflect larger ideologies. The litany of violations that occur in reality – such as the University of Oregon paying off a close associate of high school recruits to encourage them to sign with their program, Iowa State making impermissible calls beyond NCAA regulations, or the University of Tennessee colluding with a booster to provide illegal benefits to potential recruits – illustrates the myriad ways programs attempt to circumvent these rules.[19] In *NCAA Football,* however, the portrayal of recruiting as clean, smooth, and systematic masks this unsavory situation. Rather than allow programs to possibly break the rules and risk incurring NCAA sanctions, all of which would expose the Byzantine complexities of the NCAA rulebook and the Sisyphean task of trying to uphold it, the universe within the game is one where NCAA enforcement is unnecessary because any notion of breaking the rules does not exist.

For player infractions and discipline, the game's system assumes no players can ever be caught in off-field issues. Players never commit any infractions regarding team rules, university policy, or the law and subsequently never become suspended or generate negative headlines. This, of course, is not at all true on the ground in college football; players do get into trouble and oftentimes expose how universities prioritize football over model behavior. Examples abound: Lawrence Phillips had an extensive record of violent behavior, including domestic abuse, while at the University of Nebraska but received only minor suspensions as the team won the 1995 national championship with him as the starting running back. In 2013 Johnny Manziel, winner of the 2012 Heisman Trophy, was initially suspended before the season for a disorderly conduct charge, only to be reinstated by the team and Texas A&M University. Finally, the arrest of former New England Patriots tight end Aaron Hernandez for murder has created retrospective criticism of the University of Florida and his coaches while playing as a Gator.[20] These problematic scenarios are played out each time at multiple institutions, yet within *NCAA Football* they are simply swept under the rug, to be quieted under the veneer of model student-athletes as presented by the NCAA.

Moreover, the idealized simulation is reflected in the denatured representation of economic exploitation. For instance, while athletes in the game can be injured and even be lost for an entire season, they never suffer career-ending injuries. This ignores the realities of collegiate athletics, such as when Rutgers defensive tackle Eric LeGrand was paralyzed from the neck down after a brutal collision in a game against Army on October 16, 2010. By not allowing the long-term health of the athletes to be imperiled and ignoring the economic realities of collegiate sports, the class element of college football – that the sport is primarily played by athletes who tend to hail from lower-class backgrounds, yet receive no compensation outside of tuition and boarding while generating millions of dollars in revenue and media exposure for their institutions – is ignored.[21]

Finally, when it comes to academics, systematic failures within elite NCAA college football programs are masked. Within the EA game, every athlete whose eligibility runs out automatically "graduates" from their school, even though the most recent figure for graduation rates at the Football Bowl Subdivision level is at 70 percent.[22] Implicit here is the NCAA's mantra that its student-athletes "go pro in something other than sports." Every football player is assumed to have received an education – a form of fair compensation for their athletic contributions – and has plenty of opportunity to succeed after his playing days have concluded.

Within the dynasty mode of *NCAA Football* exist several aspects that are indicative not of the true nature of major college football, but rather how the NCAA (and perhaps EA Sports) would wish for its audience to see it. Problematic elements involving recruiting violations, player misconduct, and economic exploitation are simply ignored in favor of presenting the facade that every program, every player, and every aspect of the sport are beyond reproach. To that end, EA Sports is still showing its users a glimpse into big-time college football, just perhaps one approved by its licensors.

WHAT'S NOT IN THE GAME: ABSENCE OF "REAL" COLLEGE FOOTBALL?

While censorship of compromising aspects of big-time NCAA football could be explained by profit margins and brand image, if we consider the absences and gaps through the lens of simulation theory we find disturbing results about an absence of "real" college football in these digital forms. In some ways, the absent elements are more "real" than what the NCAA approved for inclusion. This potentially creates a skewed understanding of college football among some of its biggest fans. Key absences occur around academics, player representation, roster management, and the financial stakes of college football.

Some elements within dynasty mode may be more indicative of the true nature of college football than initially presented. One example can be found in the treatment of academic pursuits. Rather, the complete absence of academics from the game reaffirms the reality that athletes for major Division I football programs are by no means "student-athletes." Namely, academics are not in any way a part of the recruiting process. While recruits can be scouted for athletic prowess and even be given the labels of "Hidden Gem" or "Bust" based on how well they rate, there is no means to gauge a recruit's academic aptitude. There is never a question of whether any recruit will qualify academically, or if they can stay eligible once enrolled at the university. Moreover, while in reality many prestigious academic institutions such as Duke and Stanford take into account an athlete's grade point average when recruiting him, players in *NCAA Football* can recruit any athlete at these institutions with no such restrictions. From a procedural rhetorical standpoint, academics are deemphasized from a major concern to being nonexistent.

In light of this facet, the game is more indicative of the reality of college football. Given that the term *student-athlete* was originally concocted to prevent the NCAA from being held legally liable for the on-field death of a football player, the game, perhaps ironically, reveals the emptiness of the term in major college football.[23] Academics are completely absent, and in taking on the role of a program's coach the player is encouraged to see members of the team solely as football providers and not as students. While the NCAA has a popular series of commercials featuring the slogan, "There are over four hundred thousand NCAA student-athletes, and most of us will go pro in something other than sports," within *NCAA Football* there is never any question that these recruits are brought in for the sole purpose of playing football.

Additionally, the game, through its design absences, is unable to capture how actual college football players are often evaluated by "big-time" programs. In dynasty mode the football players are provided randomly generated names, faces, and jersey numbers, and they each hail from a hometown, which has an impact on the game's formula for recruiting. Within these details, only statistical attributes regarding their athletic prowess are provided – no qualitative or personality-driven characteristics are provided. This contrasts with steady discussion within college football about the "character" of players. For instance, Stanford head coach David Shaw has frequently espoused the importance that his recruits have the attributes of "Stanford men," which includes qualitative attributes such as confidence and communication skills. Two-time national championship coach Nick Saban issues versions of the following statement when asked in the off-season about how he continues to stock his roster: "To get some of them to come here in the summer I think is a really big tool in evaluation as well as an opportunity to get to know guys, to see if they have *the right character and attitude to fit in your program.*"[24] While EA attempts to quantify certain abstract concepts – such as the ability to recognize a play – other aspects that would point to a well-rounded student-athlete such as leadership, intelligence, and work ethic are nowhere to be found. Despite evidence from prominent coaches to the contrary, gamers are taught by EA and the NCAA to treat their players as nothing more than commodified labor.

This is also reflected in how the dynasty mode expects gamers to manage their rosters. Each season, the gamer's team is allocated only eighty-five scholarships. Yet even if the player turns off simulational elements that allow the computer to offer scholarships to recruits, the team will sign the maximum possible amount of twenty-five recruits each season. Obviously, this frequently leaves players forced to cut members of the team. The process for cutting players, however, should be noted. Conway argues that even mundane aspects of a game, such as the menu screens, can be rhetórical. *NCAA Football* certainly reflects this. After all of the recruits are signed, players then go the next step: assigning their athletes respective positions. Based on the athletes' rated skill sets, they will be proficient or deficient in particular positions. Once the game's player has assigned positions, the next stage is called "Offseason Training Results." Here, the game quantifies which athletes "improved" the most based on how many attribute points they gained. Some athletes improve at a faster rate than others, all randomized by the system. It is after these two stages that gamers are then forced to cut the roster down to eighty-five. As Baerg argues, this faceless quantifying of digital athletes into numerical categories allows players to exert control. The message from the game is clear: you cut those whose quantified skill set either was too deficient at the outset or failed to "improve" to prioritize roster space. Although perhaps not intended, this reinforces to the gamer the same mentality that currently pervades college football: athletes are treated as expendable parts. This design process actually matches up to common occurrences today, as college and university teams frequently sign over their limit and then either void existing scholarships with current team members or force recruits to delay enrollment for a semester.[25]

Finally, the role of economics is addressed – or avoided – in significant ways. There is never any report on whether a university makes or loses money from the football program each year, and a program's athletic facilities become improved simply by winning more games. While this in no way is representative of what occurs in every athletic department across the country, it conveys the notion that deficits do not matter in college football. Additionally, gamers are allowed to move teams around in any conference at will with no repercussion regarding fan or media backlash or lack of geographic continuity. Again, this illustrates how major collegiate athletics has chosen to eschew historical rivalries or geography in favor of finances and favorable television deals. In perhaps the most stark example, Rutgers University and the University of Maryland announced in 2012 that they were leaving the Big East Conference and Atlantic Coast Conference, respectively, for the Big Ten, as both schools had lost tens of millions of dollars from their football programs and could reap as much as twenty-five million dollars apiece per year in new television contracts.[26] While these financial incentives are silent within the game *NCAA Football*, the procedural options afforded gamers actually confirm the cold, calculated nature of major college football today.

DISCUSSION AND CONCLUSIONS: IDEOLOGY
AND SELECTIVE SIMULATION

From our experience with *NCAA Football*, one main observation, that it does not easily fit into established genres and typologies of sport games, will enhance the analysis to follow. More than being anomalous, we believe that the dynasty mode of this game presents a unique set of features that advance our understanding of how highly simulated video games engage with ideological systems in society.

The game does not rest easily within traditional game studies, or "ludology." In Frasca's terms, borrowed from Caillois, *NCAA Football* would be considered a "ludus" game as opposed to "paida." Paida games, reminiscent of Huizinga's early work on the subject of games, are characterized by creative play, childlike curiosity, and flexible structures. Ludus games entail rules that, in general, can easily identify a winner and loser.[27] However, these categories are broad and based on Aristotelian ideals, thus not entirely suited to the welter of game options today. Similarly, Frasca and others chafe at the older paradigms of narrative analysis of games – although EA's games have narrative elements (commentators provide contextual comments, and seasons of win-loss records provide arcs of triumph or failure), the intense detail and simulation suggest larger phenomena at work.

Conway moves beyond narrative elements to assess nondiegetic elements of games, start menus and introductory videos. In his assessment of four different digital soccer games, Conway provides a tripartite typology: televisual games (intense focus on visuals and graphics), extreme (exaggerating some aspect of the game in nonstandard ways), and managerial (focus on detailed, systematic choices). His example of a managerial game, *Football Manager 2007*, bears close resemblance to the dynasty mode of *NCAA Football* – gamers "emulate the experience of the modern football manger, creating an ideal subject position that encourages players to be meticulous, logical, and shrewd."[28] However, in *NCAA Football*, gamers also control players during the game, expanding their experience from managerial to athletic. In addition, the EA game has an intense level of visual simulation, featuring scenes of actual college stadiums, team colors, cheerleaders, and mascots. *NCAA Football*, in our view, straddles the genres of televisual and managerial.

NCAA Football is certainly procedural, especially in the dynasty mode. Working through the setup screens to assemble a roster and set the year's schedule can occupy hours of gameplay before a single snap is taken! Each of these steps is highly detailed, controlled, and systematic. There is a huge level of choice within these procedures, exemplifying how a "computer magnifies the ability to create representations of processes."[29] Consistent with the consuming level of detail in the game is the reliance on statistics and quantitative measures. Baerg notes that player-attribute systems in video games that quantify an athlete's contributions to team success put gamers in positions of control and management. He argues, "Accessing the player ratings necessarily positions gamers to exercise power over virtual athletes."[30] Despite the risk of dehumanization through quantification, the emphasis in *NCAA Football* seems to remain on coaching and managing one's dynasty: games can be played on automatic settings, or games can be canceled and replayed with no consequence. This laxity is not offered for player selections and lineups. This may be reasonable for a dynasty mode where longevity and legacy are the main goals over ephemeral in-game glory.

Another common category of digital games is the first-person shooter, most commonly seen in war simulation and other weapons-based games. Sport games can accomplish this perspective, wherein a gamer sees the field from the player's eyes, dodging or shooting or maneuvering toward his goal – but it is rare. *NCAA Football* does not offer this first-person view during game action. We offer that *NCAA Football* provides an "omniscient player" experience, as gamers occupy the positions of recruiter, head coach, offensive and defensive coordinator (entire schemes can be selected for offense and defense, which install playbooks of potential routes and coverage), as well as player (offense, defense, and special teams).[31] Gamers in *NCAA Football* all but shake the pom-poms and announce the score.

Due to this multiplicity of roles, we also cannot group *NCAA Football* with games that offer avatars and "second life" experiences for the gamer. There are elements of fantasy or vicariousness, to be sure, to the extent that gamers can choose to make a favorite team into a "dynasty," living out a long-cherished dream in that way. (One of the co-authors, a native of Minnesota, has played the Golden Gophers long enough in dynasty mode to make them overwhelming perennial champions, a far cry from the school's usual football success record.) Likewise, rather than playing a character within a complex, preplanned narrative, *NCAA Football* offers gamers, in many ways, the chance to plan their own narrative and then take part in its execution. Picking players, setting schedules, realigning conferences and bowl game associations – gamers in *NCAA Football* have a wide array of controls over the terrain of their game experience, far more so than a gamer in *World of Warcraft,* who can choose avatar, weapon, alliances, and movements but cannot, say, reshape a whole continent or rearrange tribal structures.

Other features of games that have been central to past scholarship are similarly missing from *NCAA Football.* Kingsepp's work on World War II–based historical games provides important insight into the way nostalgia for simpler, more "real" past eras is achieved in digital gaming. She notes that the troubling result of this nostalgic function is that evil-doers such as Nazis or fascists become part of a utopian reclamation of a "reality" gamers feel may be lost in current society.[32] *NCAA Football,* by contrast, has no "bad guys" and engages in no historical revisionism. Teams can lose games, to be sure, though rosters and other settings can eventually be stacked toward team success if a dynasty gamer does his or her job right. Referee calls are not a significant source of distress in the game, there is no cheating, and NCAA sanctions do not interrupt a season (more on this below). As for temporality, as mentioned earlier, the game, if anything, tinkers with future results instead of altering past events.[33]

The most significant feature of the EA game *NCAA Football,* we argue, is the presence of the NCAA. Our analysis shows that the NCAA allows only certain elements of "real" aspects of college football into the game and holds others back. From our review of gaming literature, this dynamic of third-party sponsor or authority is underexamined. Scholarship exists on "advergames," defined by Bogost as "any game created specifically to host a procedural rhetoric about the claims of a product and services," and it is tempting to consider all of *NCAA Football* as an advertisement for the NCAA.[34] However, examples of advergames are much more straightforward, with names such as *Mountain Dew Skateboarding* or *Datsun 280 Zzzap,* wherein a product is the main player or feature within the game. For EA's *NCAA Football,* there are more obvious product placements in the game, such as the recruiting service Sparqq, whose logo is displayed during games and on most screen displays during recruiting phases of the game. Past sponsors have included Coke Zero and Dove skin care for men. In the 2012 game played for this analysis, as for other years, ESPN agreed to license its logo and to provide commentators for play-by-play action during the game, but we argue that the NCAA's role in the game is different from these more obvious product promotions.

A more productive comparison may be *America's Army*, the highly simulated army warfare game developed directly by the U.S. Army to increase recruitment of enlisted soldiers. *America's Army*, often referred to as the "the official U.S. Army game," was developed by the U.S. military with the express goal of informing and interacting with U.S. popular culture. The game also functions as a recruiting tool, an admitted secondary goal of the game, especially when several versions of the game appear in highly trafficked shopping malls or other consumer destinations. The gameplay in *America's Army* is highly simulated, and while focused on army infantry roles it does allow gamers to experience "varied opportunities awaiting gamers who pursue a career in the Army." Like *NCAA Football*, not all simulations are allowed by the game's main sponsor: players are always U.S. forces, never jihadi "terrorists"; visual depiction of gore and bloodshed is nearly nonexistent, a stark departure from commercially developed war games; and even technical details such as gravity's effect on bullet trajectories were not factored in early versions of the game.[35]

However, unlike EA's college football game, *America's Army* is government owned, designed, and fully controlled. It is much closer to an advergame – with weighty and highly ideological content – than the EA product. As scholars note, there is an overt persuasive agenda to *America's Army*.[36] For EA, the NCAA is a third-party guarantor of authenticity, a bought-and-paid-for label that connotes realistic game design and gameplay. The NCAA exerts influence, eliminating elements of college football it deems unsavory, much as the U.S. Army may have decided to not show the full bloodshed of war, but it does so from a sideline position rather than the driver's seat.

We suggest that the NCAA's role is unique from profit-driven product placement advertisers and from codesign sponsors such as the U.S. Army. Assessing the NCAA's selective approval of specific elements of college football through Baudrillard's theories of simulacra and Bogost's scholarship on the simulation gap, we close by taking on the question of ideology within simulated forms. Based on our analysis, we suggest

a theory of "selective simulation" in which third-party institutions advance ideological claims via careful control over what elements of "reality" become simulated. That is, while in many ways Baudrillard's theories
could have predicted that only some of the "reality" of NCAA college
football would get into *NCAA Football*, we maintain a space for ideological critique of exactly *which* elements made the cut. Our question is
not how simulations generate new models of reality, but what reality is
allowed to be simulated in the first place?

A provocative hint toward the answer to these questions lies in the
2006 version of *NCAA Football*. The 2006 version challenged gamers
with the risk of sanctions and player infractions, asking gamers to "maintain the integrity of their program. . . . Keeping close tabs on your team
discipline will result in a clean program that attracts other outstanding
student-athletes."[37] Gamers had to choose whether to suspend players
for breaking team rules or committing other infractions or whether to
risk the ire of the NCAA for leniency. Gamers had the ability to choose
how extensively to discipline a player based on a budget of "discipline
action points" and were provided incentives for conducting a "clean" program. The NCAA demanded that EA pull the game from the shelves, and
to this day a "kill" order exists at resale stores for this version of the game.

The elements of college football present in the outlaw 2006 version
reveal the difficulty of simulated forms and ideology critique. If the
NCAA allowed the sanctions risk and discipline budget, the game would
more closely resemble the tough decisions actual football coaches and
athletic administrators make on a regular basis. However, it would hurt
its image as a pure, "amateur" athletic institution. In censoring this version of the game, the NCAA chose to present a form of simulated reality
that protected its brand image by striving to produce gamers and fans
with a skewed sense of how college football programs operate. There is
a chance, through the logic of simulacrum, that when the actual NCAA
has to punish the actual college program that you've been playing in dynasty mode, the gamer-fan would reject the "reality" that infractions and
discipline occur within sport. As Giddings reminds us, simulacral events
are still "real" – they create a strong sense of reality for participants.[38]

The 2006 version also raises other "what if" questions, namely,
the lost pedagogical opportunity of providing sport gamers with ethi-

cal dilemmas and decision-making challenges that coaches, managers, and administrators need to be well trained in navigating. As Crocco argues, digital games can become a source of social critique by offering "an alternative game-based learning with critical pedagogy to promote critical thinking." Citing scholarship that sees game-based learning as commensurate with the high-tech, collaborative, fast-paced nature of current society, Crocco focuses on how "the valuable learning principles embodied in games [can be] used to promote critical thinking about hegemonic ideas and institutions rather than to propagate them."[39] From this perspective, by censoring some of the toughest questions coaches have to make, the NCAA greatly reduced the positive impact its licensed game could have had on participants.[40] Put simply, the game as it stood in 2014 made it relatively easy to build a football dynasty by avoiding the minefields of sanctions and bad behavior. A powerful opportunity was lost to teach the complexities of building and maintaining actual sport dynasties in our contemporary social setting.

POSTSCRIPT

The selective simulation choices that the NCAA made were not without protest. Former UCLA basketball player Ed O'Bannon has spearheaded a class-action lawsuit against the NCAA, EA Sports, and the College Licensing Company—the agency that doles out NCAA licenses to approved parties—alleging that current and former collegiate players have had their likenesses copied without their consent and subsequently require compensation for video games such as *NCAA Football*.[41] The lawsuit, still working its way through federal courts, even contends that EA Sports knowingly derived their ostensibly fictive avatars from real-life players based on spreadsheets containing body types, athletic ability, and appearance. The O'Bannon lawsuit is directly cited as a primary reason that the NCAA announced in August 2013 that it would not allow its name or logo to be used in versions of the EA's college football game past the 2014 edition.

Reaction to the NCAA pulling its involvement with EA's game was initially skepticism about the impact this change would have on the video game itself. The media hype surrounding the release of *NCAA Football*

2014 attests to the game's ongoing popularity, and top digital game blog-
gers suggested that EA no longer needed the NCAA name and logo for
its blockbuster football game series.[42] However, on September 26, 2013,
EA Sports announced that the game would cease to be produced after
the 2014 version.[43] EA explicitly blamed the NCAA's ongoing legal woes
with cases such as O'Bannon's for the decision. Their official statement
read: "We have been stuck in the middle of a dispute between the NCAA
and student-athletes who seek compensation for playing college football.
Just like companies that broadcast college games and those that provide
equipment and apparel, we follow rules that are set by the NCAA – but
those rules are being challenged by some student-athletes."[44] In October 2013, EA reached a $40 million settlement with the athletes in the
O'Bannon case.[45] The NCAA remains the sole defendant in the case, due
for trial in June 2014.

Despite these decisions and pending litigation, scholars interested
in sport, gaming, and social theory would be wise to continue studying
these powerful cultural forms. Additional methodological approaches
are always valuable in this endeavor. Kingsepp's admonition, for instance, that "practical gaming experience does not automatically confirm [theoretical conclusions]" is valuable. Ethnographic research with
NCAA Football gamers would help us assess the fourth level of Baudrillard's scale, the extent to which simulacra are produced by the selective
simulation presented in games such as EA's college football example.
We maintain, however, that the question of ideological process within
simulated forms has merit and can be pursued. As Cubitt concludes,
"Communication is not just about the relationship with the real but
about relationship with other people."[46] Powerful, popular simulated
games such as *NCAA Football* extend communication between human
agents and subjects, with the very real possibility of ideological effects.
Tracking what reality is allowed to be simulated, and to what effect, is
valuable and important.

NOTES

1. Aaron Suttles, "Game On: With
New College Season Almost Here, EA

Sports Releases *NCAA Football 14*," *Tuscaloosa (AL) News*, July 9, 2013, C1.

2. Brett Molina, "*NCAA Football* 12 Breaks Franchise Record," *USA Today*, July 26, 2011, http://content.usatoday.com/communities/gamehunters/post/2011/07/ncaa-football-12-breaks-franchise-sales-record/1.

3. "College Sports: Fair or Foul?," *Economist*, April 27, 2013, http://www.economist.com/news/united-states/21576688-unpaid-student-athletes-are-heart-multi-billion-dollar-industry-fair-or-foul.

4. Molina, "*NCAA Football* 12 Breaks Franchise Record."

5. See Sean Cubitt, *Simulation and Social Theory* (London: Sage, 2001).

6. Epsen Aarseth, *Cybertext: Perspectives on Ergodic Literature* (Baltimore: Johns Hopkins University Press, 1997); Gonzalo Frasca, "Simulation versus Narrative: Introduction to Ludology," in *Video/Game Theory*, edited by Mark J. P. Wolf and Bernard Perron (New York: Routledge, 2003), 221–235; Seth Giddings, "Dionysiac Machines: Videogames and the Triumph of the Simulacrum," *Convergence* 13 (2007): 417–431.

7. Frasca, "Simulation versus Narrative," 223.

8. Giddings, "Dionysiac Machines," 421.

9. Jean Baudrillard, *Simulacra and Simulation* (Ann Arbor: University of Michigan Press, 1994), 1–2; Cubitt, *Simulation and Social Theory*, 50.

10. Baudrillard, *Simulacra and Simulation*, 41n7; Cubitt, *Simulation and Social Theory*, 42.

11. David J. Leonard, "Live in the World, Play in Ours: Race, Video Games, and Consuming the Other," *Studies in Media and Information Literacy Education* 3, no. 3 (2003): 1; David J. Leonard, "It's Gotta Be the Body: Race, Commodity, and Surveillance of Contemporary Black Athletes," in *Studies in Symbolic Interaction*, edited by Norman Denzin (Bingley, UK: Emerald, 2009), 33:180; Dean Chan, "Playing with Race: The Ethics of Racialized Representation in E-Games," *International Review of Information Ethics* 4 (2003): 25.

12. See Leonard, "Live in the World, Play in Ours."

13. Frasca, "Simulation versus Narrative," 232.

14. Ian Bogost, *Persuasive Games* (Cambridge: MIT Press, 2007), 5, 3; Andrew Baerg, "Governmentality, Neoliberalism, and the Digital Game," *Symploke* 17, nos. 1–2 (2009): 115–127; Bogost, *Persuasive Games*, 25.

15. To a large degree, Bogost, *Persuasive Games*, makes a similar claim, but focused on "ideological frames" provided by procedures in the areas of politics, advertising, and education.

16. Ibid., 43.

17. Steven C. Conway, "Starting at 'Start': An Exploration of the Nondiegetic in Soccer Video Games," *Sociology of Sport Journal* 26 (2009): 69; Bogost, *Persuasive Games*, 29.

18. We maintain a focus on the NCAA here, despite its termination of its EA contract, since the elements assessed are qualitative aspects of gameplay, not details of team or stadium licensing. How these elements may shift after the 2014 version of the game has yet to be seen.

19. See Steve Eder, "Ducks and Former Coach Punished by the NCAA," *New York Times*, June 27, 2013, B12; Randy Peterson, "Iowa State Football, Basketball Guilty of Major NCAA Violations," *USA Today*, April 12, 2013, http://www.usatoday.com/story/sports/college/2013/04/11/iowa-state-found-guilty-major-ncaa-violations/2074593/; Gary Klein, "Lane Kiffin Claims No Knowledge of Alleged Violation," *Los Angeles Times*, September 27, 2011, http://articles.latimes.com/2011

/sep/27/sports/la-sp-0928-usc-garza
-lyles-20110928.

20. Willie McDuffy, "The Mysterious Case of Nebraska's Lawrence Phillips," *Bleacher Report,* October 2, 2010, http://bleacherreport.com/articles/480074 -mysterious-case-of-lawrence-phillips; S. C. Gwynne, "How Johnny Football Almost Didn't Play for Texas A&M," *Texas Monthly,* June 19, 2013, http://www.texas monthly.com/story/how-johnny-football -almost-didnt-play-texas-am; Mike Bianchi, "Has Aaron Hernandez Supplanted Tim Tebow as Symbol of Urban Meyer's Gators?," *Orlando Sentinel,* July 2, 2013, http://articles.orlandosentinel.com/2013 -07-02/sports/os-mike-bianchi-urban -meyer-aaron-hernandez-0703-20130702.

21. David Epstein, "My Sportsman: Eric LeGrand," *Sports Illustrated,* December 1, 2011, http://sportsillustrated.cnn .com/2011/magazine/sportsman/11/28 /ericlegrand.sportsman/index.html; "College Sports: Fair or Foul?"

22. George Schroeder, "Notre Dame, SEC Also Win in Classroom," *USA Today,* October 25, 2012, http://www.usatoday .com/story/sports/ncaaf/2012/10/25/ncaa -graduation-rates-notre-dame-sec-college -football/1656329/.

23. Taylor Branch, "The Shame of College Sports," *Atlantic Monthly,* October 2011, http://www.theatlantic.com /magazine/archive/2011/10/the-shame-of -college-sports/308643/.

24. Dan Wetzel, "David Shaw Building Stanford into a Perennial Power with Unconventional Recruiting Mandate," *Yahoo! Sports,* March 27, 2013, http://sports.yahoo .com/news/ncaaf–david-shaw-building -stanford-into-a-perennial-power-with -unconventional-recruiting-mandate -044557382.html; Mike Herndon, "Nick Saban's Final Crimson Caravan Message: 'The Only One That Matters Is the Next

One,'" May 22, 2013, http://www.al.com /sports/index.ssf/2013/05/nick_sabans _final_crimson_cara.html.

25. Conway, "Starting at 'Start'"; Andrew Baerg, "Classifying the Digital Athletic Body: Assessing the Implications of the Player-Attribute-Rating System in Sports Video Games," *International Journal of Sport Communication* 4 (2011): 133–147; Bernard J. Machen, "Florida President: Grayshirting Is Morally Reprehensible Practice," *Sports Illustrated,* February 1, 2011, http://sportsillustrated.cnn.com /2011/football/ncaa/01/31/bernard .machen.letter/index.html.

26. Patrick Rishe, "Win-Win as Rutgers, Maryland, Big Ten Cash In," *Forbes,* November 20, 2012, http://www.forbes .com/sites/prishe/2012/11/20/win-win -win-as-maryland-rutgers-big-ten-cash-in/.

27. Frasca, "Simulation versus Narrative," 229. See also Johan Huizinga, *Homo Ludens: Man the Player* (Boston: Beacon Press, 1955).

28. Conway, "Starting at 'Start,'" 83.

29. Bogost, *Persuasive Games,* 5.

30. Andrew Baerg, "Classifying the Digital Athletic Body," 144.

31. For instance, in publicity for *NCAA Football 14,* Suttles notes that "this year's game incorporates new gameplay elements to create a more real-life feel to the game, including . . . an improved running game with added choices for the option offense. The aim was to make the players in the game react as a real college football player would on a Saturday afternoon." Suttles, "Game On," C1.

32. Eva Kingsepp, "Fighting Hyperreality with Hyperreality: History and Death in WWII Digital Games," *Games and Culture* 2 (2007): 371.

33. A significant and newsworthy change to *NCAA Football 14* is the introduction of the "Ultimate Team" mode of

the game, which allows gamers to construct "all-time" teams from more than fourteen hundred former college football greats, who have given (or sold) the rights to use their names and likenesses, as well as career stats, to give gamers the chance to shape multigenerational teams of ultimate talent. See Suttles, "Game On," C1.

34. Bogost, *Persuasive Games.*

35. David Neiborg, "Training Recruits and Conditioning Youth: The Soft Power of Military Games," in *Joystick Soldiers: The Politics of Play in Military Video Games,* edited by Nina Huntemann and Matthew Thomas Payne (New York: Routledge, 2010), 54, 61; David Nieborg, "America's Army: More than a Game," in *Transforming Knowledge into Action through Gaming and Simulation Conference,* proceedings from 2004 ISAGA, edited by Thomas Eberle and Willy Christian Kriz, http://www.gamespace.nl/content/ISAGA_Nieborg.PDF.

36. Neiborg, "Training Recruits and Conditioning Youth."

37. Insight into the 2006 version was gained from co-author experience and correspondence with Owen Good, columnist for *Kotaku, the Gamer's Guide.* Good stated that since the NCAA banned the version, it has been "killed" from resale shelves. One of the co-authors purchased a version via an internet website and played it extensively. When playing the game, random warning notices pop up, announcing that "one or more user controlled teams have players on their team that have committed NCAA infractions." The gamer then chooses between "Take action" and "Continue with no action." Additional screens specify the player and type of infraction – falsified paperwork, for instance – and report the gamer's "Discipline Action Points Left" for dealing with current and future issues. Good reports that

the prior version also included celebration routines not allowed in the current release, as well as girlfriends (at least in the version where gamers control one player in pursuit of a Heisman Trophy). Good has written about this past version at http://kotaku.com/5977523/fake-girlfriends-and-heisman-glory-are-nothing-new-to-video-games.

38. Giddings, "Dionysiac Machines," 429.

39. Francesco Crocco, "Critical Gaming Pedagogy," *Radical Teacher* 91 (2011): 27.

40. E-mail correspondence with Owen Good suggests that a prior version of the game mode when gamers controlled one player in pursuit of a Heisman Trophy included features for gamers to take their player to class and take tests. This feature has since been removed; Good offered no information as to who made this decision.

41. Steve Berkowitz, "Documents: Electronic Arts Mimicked NCAA Athlete Likenesses," *USA Today,* June 20, 2013, http://www.usatoday.com/story/sports/college/2013/06/20/obannon-vs-ncaa-electronic-arts-clc-lawsuit-hearing/2440481/.

42. One day after the NCAA announced the end of its EA relationship, Good reported information from a source knowledgeable about the case that EA had paid the half-million-dollar licensing fee to the NCAA to gain access to its college basketball material, namely, the Final Four tournament images. Good suggests that EA didn't need the NCAA symbols for its football game and may have been relieved to be free of a hefty and unnecessary price tag. See http://kotaku.com/ea-sports-didnt-need-the-ncaas-logo-and-maybe-it-did-860124604. See also http://kotaku.com/the-ncaa-will-not-renew-its-licensing-agreement-with-ea-815771708.

43. Tony Manfred, "EA Sports Cancels Its College Football Video Game amid a Wave of Lawsuits," September 26, 2013, http://www.businessinsider.com/ea -sports-cancels-ncaa-football-videogame -2013–9#ixzz2wNONcNqo.

44. Ibid.

45. Samit Sarkar, "Lawyers Never Intended for EA to Stop Making NCAA Football Games," October 5, 2013, http:// www.polygon.com /2013/10/5/4803216/ea -sports-ncaa-football-cancellation -lawyers-never-intended.

46. Kingsepp, "Fighting Hyperreality with Hyperreality," 373; Cubitt, *Simulation and Social Theory,* 152.

TEN

Yes Wii Can or Can Wii? Theorizing the Possibilities of Video Games as Health Disparity Intervention

David J. Leonard, Sarah Ullrich-French, and Thomas G. Power

THE DEBATE ABOUT EXERGAMING OFTEN APPEARS IN headlines such as "Can Wii Games Replace Regular Exercise?" and "Is the Wii Fit Better than Regular Exercise?"[1] In this regard, virtual gaming has been reduced to a binary, a mathematical formula that treats participants as universal subjects and analyzes how well the games transport those bodies into virtual space. It reflects on whether these games have real-life impact on the universal game subject and how these virtual activities compare to their real-life brethren. Take one study from the American Council on Exercise, which after testing sixteen participants on six of Wii's most challenging games – Free Run, Island Run, Free Step, Advanced Step, Super Hula Hoop, and Rhythm Boxing – concluded that virtual reality was distinctively different from the real world, in that twice as many calories were burned with the real "thing." Emblematic of much of the discourse, the adherence to the virtual-real binary and its conceptualization of all participants as having equal access and opportunity demonstrate the shortcomings of the discourse surrounding virtual exercise.[2] Furthering the establishment of this dualistic framework, the discourse focuses on the caloric impact–energy expenditure rates of virtual exercise games; it works to understand if exergaming is a substitute for real-world exercise. Yet there has been little effort to measure the impact of games on the physical body (core strength, balance) and, more important, the impact of games on identity, knowledge about fitness, health, and nutrition. In the end, these studies, more than the games themselves, disembody people and fail to look at how games change people in a myriad of ways, from the physical to the mental, from identity to self-worth.

Cyberculture discourse constructs barriers, obstacles, and other impediments between cyber pleasures and real-life needs. "For couch potatoes, video game addicts and surrogate travelers of cyberspace, an organic body just gets in the way," writes Margaret Morse. Similarly, Deborah Lupton highlights this entrenched cyber-real divide, noting, "In cyberwriting the body is often referred to as 'the meat,' the dead flesh that surround the active mind which constitutes the 'authentic self.'" Virtual reality offers people "the dream . . . to leave the 'meat' behind and become distilled in a clean, pure uncontaminated relationship with computer technology."[3] Evident here, and elsewhere, the public and academic discourses ubiquitously conceptualize virtual reality at the interface between the mind and the represented space. Technology provides a vehicle for imagination, simulation, and connectivity between mind and machine. The body is superfluous. Whereas the virtual space exists for the mind, the real world serves the body – it is the space for "the meat." Exergaming presents itself as the opportunity to bridge the divide by bringing "the meat" to the cyber world. It provides a space, a virtual landscape, that brings the body into play.

According to McKenzie Wark, a virtual landscape is a "terrain created by the television, the telephone, the telecommunications networks crisscrossing the globe." In *Virtual Geography: Living with Global Media Events,* Wark argues:

> These "vectors" produce in us a new kind of experience, the experience of "telesthesia" – perception at a distance. This is our "virtual geography," the experience of which doubles, troubles, and generally permeates our experience of the space we experience firsthand. This virtual geography is no more or less "real." It is a different kind of perception, of things not bounded by rules of proximity, of "being there." If virtual reality is about technologies which increase the "bandwidth" of our sensory experience of mediated and constructed images, then virtual geography is the dialectically opposite pole and process. It is the expanded terrain from which experience may be instantly drawn.[4]

Though not privileging the real over the virtual, the tendency evident here is to construct walls between what is simulated and what is lived, what is visual and what is physical, and what is connected to mind and what interfaces with body.

This chapter seeks to reflect on the illusion of this binary in multiple ways. We contemplate the dialects between virtual exercise and "the meat" at work during exercise games. Likewise, we also use this space to reflect on how real inequalities, whether visible with differential access to safe physical activities or the availability of nutritious food, to further break down these barriers. In pushing the discussion toward simulation, embodiment, and the dialects between real and virtual, we use this space to highlight the potential usefulness of exergames in addressing lived inequalities while also noting the limitations in these games because of the lived inequalities.

Likewise, this chapter looks to examine video games as a potential source of change and transformation in the battle against health disparities and inequalities. In examining rates of video game usage among youth of color, health disparities, the digital divide, and the games themselves, this chapter points to the possibilities and limitations of the virtual fit movement within poor communities of color. From the ways that "choice" is framed and the emphasis on engaging in physical activity inside and outside of virtual reality to the ways in which games transform identities and open up virtual geographic spaces, this chapter thinks through the ways in which the virtual fit movement can challenge inequalities even among the replication of inequalities within the games themselves.

VIDEO GAMES

In recent years, much of the public discourse and some scholarly research have purported a link between video game usage among kids and rates of obesity.[5] Given that the average kid ages eight to twelve spends thirteen hours per week playing video games, and given that a recent study found that 9 percent of American youth are clinically and pathologically addicted to games with another 23 percent confessing that at times they have felt addicted to games, it is no wonder that video games have elicited concerns about childhood health and obesity.[6]

Responding to the obesity epidemic plaguing Western industrial nations, the video game industry turned its attention to games that advanced an agenda of health and exercise. In 2006 Casual Living released

Yourself Fitness, a game that merged together gaming technologies with the exercise DVD market, providing players with fitness routines, meal planning tips, and exercise advice that could be done in one's home. Akin to exercise DVDs, *Yourself Fitness* incorporated the virtual aesthetics of video game culture through keeping "workouts fresh by doing them in the Empress' Dojo, the Alpine Retreat, an Island Paradise, the Urban Oasis, the Desert Springs Resort or a Meditation Garden."

The effort to use video games as part of an exercise program is nothing new. *Dance Dance Revolution* has long been a popular workout. Beginning in the 1990s, the media celebrated *D.D.R.* as a weight-loss phenom, citing ample examples as to the potential panacea of exergaming to the global obesity epidemic. Seth Schiesel, in "P.E. Classes Turn to Video Game That Works Legs," described the trend in the following way:

> It is a scene being repeated across the country as schools deploy the blood-pumping video game *Dance Dance Revolution* as the latest weapon in the nation's battle against the epidemic of childhood obesity. While traditional video games are often criticized for contributing to the expanding waistlines of the nation's children, at least several hundred schools in at least 10 states are now using *Dance Dance Revolution,* or *D.D.R.*, as a regular part of their physical education curriculum. . . . Incorporating *D.D.R.* into gym class is part of a general shift in physical education, with school districts de-emphasizing traditional sports in favor of less competitive activities.[7]

Writing about West Virginia's plan to incorporate *D.D.R.* into its physical education curriculum, Stephen Totilo similarly described the turn in gaming as one that responded to both the obesity epidemic and America's turn away from competitive sports.[8] The discursive focus was not simply about video game technology as an instrument of fitness, weight loss, or even calories burned, but a vehicle to reimagine exercise in a way that reaches all students, not simply those who find reward and fulfillment in organized sports. This would only increase with the advent of the Wii, which formally brought the body into the game.

IT'S A WII WORLD, AND WE JUST LIVE IN IT

In 2006 the potential in exergaming changed dramatically with the advent of Wii. Building upon these past games, the Wii sought to break down the barriers between the real and the virtual, utilizing the balance board and other technologies to bridge the virtual-real divide, fostering a connection – in movement, identity, space, place – between player (person) and avatar (played). Illustrating a paradigm shift, Wii created a space that both replicates and produces kinesthetic motion, thereby producing a mirror simulation of sorts. The integration of the camera, sensor bar, and balance board, previously unavailable within exergames, produced a level of integration that seemingly changed the relationship between self and avatar. You were now your avatar rather than merely simulated by said avatar, which in turn reduces the distance between the "virtual" and the "real."

Nintendo increasingly focused not simply on virtual exercise as a replacement for what could be done on the tennis and basketball courts or in the gym, but the dialectics and complementary nature of the real and the hyperreal. "We hope that Wii will encourage users to be more physically active as well as spark a discussion about fitness in the household," explained Marc Franklin, director of public relations for Nintendo of America.[9] Equally important, Nintendo sought to market the Wii not as a substitute for real-life exercise but as a tool of education, inspiration, and reorientation, assuming that its players will have an equal chance of applying those lessons in the real world. The incorporation into school settings has similarly focused on its relationship to outdoor sports, so much so that exergaming is imagined as a tool to encourage and train kids to enjoy physical activity. The emphasis is overwhelmingly on the bridge between the virtual and the real, with exergaming serving as a tool to facilitate activities in both places.

Wii Fit as exercise program allows players to partake in a number of exercises, but it also provides information about weight, body mass index, and "*Wii Fit* Age." The game comes with a virtual trainer who provides instructions and encouragement throughout the workout. The majority of the exercises take place on a balance board that not only allows the system to track player movement and weight management progress, but also challenges players to improve balance and overall fitness. Specifically, *Wii Fit* offers players more than forty different workout games, including yoga, strength training, aerobics, and even running, as well as several balance games, including a ski slalom run, a ski jump, and a soccer game. Prior to the release of *Wii Fit*, players found exercise possibilities with *Wii Sports,* a series of games that require body movement. According to the American Council on Exercise, people on average burn eight calories a minute playing "real" tennis, compared to five calories a minute playing Wii tennis. Likewise, boxing via Wii results in on average expenditure of seven calories per minute, which, although less than actual sparring, still constitutes a worthwhile workout. Bryan Haddock, a professor of kinesiology at California State San Bernardino, found that college students "use an average of 5.5 to 7.5 calories a minute" doing games on the Wii.[10]

The release of *Wii Sports* and *Wii Fit* and the widespread speculation and commentary about the potential health benefits of playing video games have prompted widespread debate within the exercise and health communities. The popularity of these games and programs among exercise gamers prompted the release of a number of games, including *Wii Fit Plus, Wii Sports Resort, EA Sports Active, Walk It Out, The Biggest Loser, Gold's Gym Cardio Workout, Jillian Michaels Fitness Ultimatum 2010,* and countless others as well as various tools (Wii dumbbells). Notwithstanding the virtual exercise craze and the efforts to make virtual exercise look like the real thing, many questioned the efficacy and usefulness of a virtual-based exercise regimen. Joseph Donnelly, professor of health sport and exercise science at the University of Kansas, expressed doubt about the usefulness of *Wii Fit* and *Wii Sports:* "Electronic gimmicks do not appear to be the solution for the physical inactivity problem that we have in this country. You're not going to be able to play a Wii game for 15 or 20 or even 30 minutes and get the kind of energy expenditure that is needed to be fit and healthy." Likewise, Michael Torchia, a fitness expert who is also organizing a class-action suit against Nintendo, concluded that *Wii Fit* and *Wii Sports* were not a solution but rather part of the problem: "Nintendo is contributing to the epidemic of obesity. Young and old are putting away their gym clothes and shying away from going outdoors to play sports, because of the addictive appeal to the Wii game products." Comparing Nintendo to tobacco companies, Torchia accused Nintendo of creating "a false image of their products" that "hid the potential dangers."[11] Yet the criticism has gone little beyond the question of calories and energy expenditure and has focused on whether exergaming replaces traditional exercise or promotes decreased health and fitness.

Presently, there have been only a couple of substantive studies to explore the usefulness of virtual exercise, albeit on relatively small scales.[12] In 2007 the Research Institute for Sport and Exercise, located in Liverpool, England, published a study that looked at the potential health benefits of playing sports games on Nintendo's Wii. Comparing the energy expenditure of adolescents who played both the more traditional sedentary video games and boxing, tennis, and bowling on the Wii sport systems, these scholars found that both energy expenditure and heart rate increased by playing Wii (boxing was found to be the best "workout"). Although the study concluded that a virtual workout from the Wii did not compare to the "real thing," the author saw potential benefits within these games. While "active gaming . . . was not intense enough to contribute toward the recommended amount of daily physical activity for children," the "new generation computer stimulated positive activity behaviors – the children were on their feet and they moved in all directions while performing basic motor control and fundamental movement skills that were not evident during seated gaming."[13] In other words, there is some evidence to suggest a small number of games can produce moderate levels of activity, but overall there is little evidence to say it should substitute for "real" activity. This study points to the possible usefulness of games, although its focus on energy expenditure (using a monitor that didn't detect arm movement) rather than weight loss (BMI), its use of a laboratory setting, the absence of an educational component, and its focus on *Wii Sports* as opposed to *Wii Fit* all point to future possibilities. The American Council on Exercise, along with the University of Wisconsin, pointed to similar possibilities with Wii, concluding that though not a viable "substitute for the real sport," "playing Wii Sports increases heart rate, maximum oxygen intake and perceived exertion, which ultimately translates to calories burned."[14]

In another study, Lorraine Lanningham-Foster found that video games that required movement have some usefulness. The researchers tested both groups of kids sitting and watching television, watching television while walking on a treadmill, playing a traditional sedentary video game, and playing two different activity-based video games. "The results showed that sitting while watching television and playing traditional video games expended the same amount of energy. When participants played with the first activity-oriented video game, one that uses a camera to virtually 'place' them in the game where they catch balls and other objects, their energy expenditure tripled." It found that with dance games, the kids burned the most amounts of calories; it also concluded that with this type of game, unlike the others, there was a discrepancy between the groups of kids, with the mildly obese kids burning more calories. The researchers thus concluded that given the amount of television watched and video games played by kids, activity-based games are certainly preferable to traditional screen time. Lanningham-Foster, an expert on obesity, described the results in the following way: "We know if kids play video games that require movement, they burn more energy than they would while sitting and playing traditional screen games. That's pretty obvious even without our data. The point is that children – very focused on screen games – can be made healthier if activity is a required part of the game."[15]

In a study that appeared in the *Archives of Pediatrics and Adolescent Medicine*, Bruce Bailey and Kyle McInnis concluded that interactive games "compared favorably with walking on a treadmill at three miles per hour, with four out of the six activities resulting in higher energy expenditure."[16] In their research, thirty-nine middle school kids played six different types of games, illustrating the varied potential of each. Specifically, participants showed a metabolic equivalent task value as follows:

- Wii: 4.2 METS
- Dance Dance Revolution: 5.4 METS
- Cybex Trazer: 5.9 METS
- LightSpace: 6.4 METS
- Xavix: 7.0 METS
- Sportwall: 7.1 METS

In addition to these numeric findings, the study determined that boys found greater enjoyment in these games and that those with the highest body mass index also showed the greatest interest in exercise games.[17]

Similarly, Scott Owens, a University of Mississippi associate professor, launched a study in 2008, exploring the potential usefulness of the Wii in facilitating improved family fitness. The research looked at two groups, where one group of subjects used the Wii system and the other lacked a Wii in the home. It looked at the fitness levels of eight families over six months.[18] The study focused on the impact of the Wii on each family's physical fitness, based on a series of different measurements, which included balance, body composition, physical activity, and aerobic fitness. These measurements, along with the data generated by the games themselves (weight, BMI, Wii fitness age), would be used in comparison to fitness levels, which the study measured prior to the start of the research.

In 2009 a Nintendo-funded study concluded that more than 30 percent of games and activities available on *Wii Fit* and *Wii Sports* "require an energy expenditure of 3.0 METs or above." The study's chief investigator, an adviser for *Wii Fitness Plus* and head of a physical activity program at the National Institute of Health and Nutrition in Tokyo, Motohiko Miyachi, hypothesized that additional time in the virtual gym "may contribute to prevention of cardiovascular diseases," even if the study didn't reveal the direct health improvements of exergaming. Examining the impact on exergaming on twelve men and women, ages twenty-five to forty-four, this study focused on the fitness potential of virtual reality. At the beginning of the study, researchers tested the heart rates and oxygen levels of its participants while also establishing a baseline of "perceived exertion." Providing its participants with coaching, the researchers taught them how to play the games, including Advanced Step, Super Hula Hoop, Free Run, Island Run, Free Step, and Rhythm Boxing, each of which they would play in six-minute intervals. What the researchers found demonstrated the potential of the games:

> Of the list of *Wii Sports* and *Wii Fit* games tested for energy expenditure and exercise effectiveness, the majority of games were considered moderate intensity exercises with ratings between three and six METs. The game with the highest MET rating was the Single Arm Stand, with a rating of almost six METs. Island Run and Free Run also ranked among more moderate intensity exercises, burning an average of 5.5 kilocalories (kcal) per minute or 165 calories on average when played for 30 minutes. The other four games selected for testing typically burned about 3.3 to 3.8 kcal per minute or anywhere from 99 to 114 calories on average for a 30-minute session.[19]

Similarly, J. C. Nitz and colleagues found that *Wii Fit* was useful in "improv[ing] balance, strength, flexibility, fitness and general well-being" for middle-aged women.[20] Facilitating an intervention twice a week for ten weeks, the study required women to engage in *Wii Fit* activities in thirty-minute sessions. While not showing significant improvement in all facets, the research revealed the potential of the Wii in improving balance and lower-leg muscle strength. According to researchers, the aerobic activities and the balance games available on Wii required participants to engage in moderate-intensity activity, akin in energy expenditure to "dancing, volleyball, bicycling, jogging or water aerobics."[21] The emphasis on laboratory studies, and the erasure of class, race, geography, and other variables that inhibit health equality, illustrates the ways in which much of the literature has imagined games as mere simulation of reality rather than an interface with reality, one that cannot transcend the real inequalities that divide society. Likewise, the focus on calorie expenditure and the ability of games to mimic real-life activities presumes a level of equity and equality between subjects. As game players exist in a world stratified by race, class, gender, and geography, the treatment of exergamers as a homogenous body represents a major shortcoming.

EXERGAMES: REAL-LIFE OBSTACLES

Exergames do have potential. Yet this potential is constrained by the real-world divisions, inequalities, and obstacles. While much of the literature focuses on the potential and pitfalls of exergames as they relate to the technology, it is crucial to reflect on how the "real" remains the truest obstacle. *The Biggest Loser* replicates many of the traditional elements of exergaming: it offers participants the opportunity to complete a myriad of exercises and activities that work strength, balance, and cardio. In an attempt to capitalize on the popularity of the television show, the Wii game not only contains the real-life contestants, whom gamers compete as and against, but also allows exergamers to participate in simulated competitions. Whereas many Wii games imagine players through their Mii characters, simulated cartoon avatars that signify the weight and height of their real-life brethren, *The Biggest Loser* utilizes simulated representations of actual contestants akin to sports video games.

In addition to prompting players to partake in yoga, aerobic activities, and other exercises, *The Biggest Loser* provides gamers with advice about nutrition, going as far as to give players recipes from blueberry pancakes to penne pasta with vegetables. The emphasis on proper diet, on caloric intake (players are asked to enter daily calories, and success in the game is actually measured by one's success in not exceeding a daily calorie limit), demonstrates the ways in which the game breaks down the virtual and the real. While exercise takes place through a player-avatar dialectic, the game's efforts to impact outside of the virtual highlight this complex relationship. *The Biggest Loser* works hard to blur the lines between the game-avatar and the physical body. It's tagline, "Real fun, real results," elucidates the manner in which the game seeks to emphasize its connectivity – the realness of the game. It is a mediation of a real physical body that is pushed to move, which in turn will result in the transformation of the real body. This results in the transformation of the virtually signified body – the transformation of the signified illustrates the real transformation, producing a circular euphoria that encapsulates the impact of the work. In game design, exergames work to create a mirror of real activities and motion rather than a simulation of reality. And while the physical body is not virtually transported into a real gym, onto treadmills, hanging above pools, or onto the beaches of Hawaii or the mountains of Malibu, the sweat, the physical movement, and the post-workout soreness are real.

This game, and countless other exergames, works to break down the walls between the virtual and real. The advice to eat sugar-free pudding (after playing the game and not in the game), to watch calorie consumption for yourself (and not your avatar), to cook blueberry pancakes from *The Biggest Loser* game recipes (in a real kitchen), and to be active in one's daily life illustrates the game's relation to the player's physical space and existence. The interface between the real and the virtual is evident through the dialectics that exist between the game and people's own attitudes. For example, here are some comments posted on an online review of the game:

> I love the biggest loser Wii game. I have lost 2 lbs. in the first week. It is definitely a challenge. I have a gym membership, but it is so hard for me to find the time to get to the gym. I always feel so rushed for time getting home from work at 6pm, cooking supper, clothes, etc. working out at home with the BL wii game is perfect. My husband & I both do it.[22]

> [H]ave the biggest loser wii and so far i've lost 11 pounds in 3 weeks i like the feeling of my b/fs money is going to use. it makes me more and more happy every week when i weight myself and see results. it seems weird to say it but the biggest

loser wii is more usefull then going to the gym. when i was going to the gym i
didnt lose anything then i did the biggest loser wii i started to see results. i think
its the fact is it works out everything and they tell you what to do. unlike going to
the gym you have no clue unless you work with a personal trainer and thats more
money your spending so yes the wii biggest loser is much cheaper and affective
then going to the gym.[23]

Rather than comparing virtual exercise to the real, researchers must
consider how and why people would opt to exercise via the Wii and not
in the "real world." Whether because of identity, class, or access issues,
some people engage in exercise through a virtual platform and not those
deemed to be in reality.

Other games also work to disentangle the assumed barriers that exist between the real and the virtual all the while creating conditions more
enticing and inspiring for a productive workout. With *EA Sports Active
NFL Training Camp,* players are able to go through a series of seventy
football-related "drills and challenges designed to improve strength,
power and conditioning, as well as reaction skills, agility and first step
quickness" within their favorite stadiums. From the privacy of their own
homes, they are able to train with top NFL players. In many regards, the
power and allure of games that transport players outside of their own
reality reflect not only the exotic and novel aesthetics available in these
spaces but the more sterile and mundane spaces available in everyday
life. Worse yet, the attractiveness and power of virtually visiting a modern gym, a yoga studio, a pristine sports complex, and even an NFL
training camp stem from the lack of access to these spaces and places,
especially when considering the isolation of America's poorest communities and communities of color.

The connection between the real and virtual transcends the connections between avatar and self, or even the ways in which the camera
and sensor work to reproduce self in virtual reality, but is evident in
the cultural, political, and social meaning of the games themselves. The
popularity of exergames is the result of what is happening in the real
world: high levels of obesity. The integration into public schools not only
reflects the obesity epidemic but the impact of No Child Left Behind
(schools needing to focus more resources and time on math and reading)
and budgetary shortfalls commonplace after 2008.

The real world exists inside of virtual exergaming. In a society where playing sports is seen as a boys activity, and where girls learn very early that their success and value are not through participation in sports, the space provided by virtual exercise is one of great potential. Similarly, the power and potential of the games stem from the ways in which health and fitness are defined through middle-class white identities. The ways in which running, yoga, and exercise in general are imagined through whiteness (and in some cases defined through a middle-class aesthetic) illustrate the potential here. While to date the Wii is far more popular within middle-class families (giving one pause as to whether Wii merely replicates the class and racial fissures of the exercise industry as a whole), it points to the connections between the real and the virtual.[24] It is not surprising that a joint study between the University of New Mexico and the Department of Agriculture "offered children and adults the opportunity to play active games at a Laundromat in Hawaii, an after-school program in Connecticut and a low-income community program in Delaware."[25] The games are imagined as real-life interventions. In England schools have used Wii as a tool to reorient students otherwise disengaged from and uninterested in physical education classes.[26] The implementation of exergame workout programs is about reaching children and adults left behind by the fitness revolution. Yet in the end, the real-world inequalities that limit fitness and health choices along racial and class lines play out in the virtuality of fitness.

CONCLUSION: EXERGAMES AS FAILED INTERVENTION

The public discourse surrounding obesity is so often defined by individuality, choices, and the notion of meritocracy. Just as games function as a space to divide the successful from the failed, just as games measure the quality of one's ability, obesity versus health is so often constructed as a referendum on one's choices, character, and ability. In reducing the issue to choice and individual character – moralism – much of the national discourse on obesity systematically erases fissures, divisions, and inequalities. This is particularly evident in popular culture, as evidenced by the widespread presence within certain reality television shows and the narrative offerings within exergaming. The cultural productions "borrow a basic set of assumptions from 'The Biggest Loser,'" writes Aaron Barnhart. "Obesity is largely a product of inertia, of spending too much time sitting around eating terrible food. The cure, therefore, is activity – lots of it, with occasional breaks to make healthy meals and visits to the confession-cam."[27] David Grazian, in "Neoliberalism and the Realities of Reality Television," describes *The Biggest Loser* as yet another reality-based show that pivots on the tenets of neoliberalism:

> Although the very design of competitive reality programs . . . guarantees that nearly all players must lose, such shows inevitably emphasize the moral failings of each contestant just before they are deposed. In such instances, the contributions of neoliberal federal policy to increased health disparities in the U.S. – notably the continued lack of affordable and universal health care, and cutbacks in welfare payments to indigent mothers and their children – are ignored in favor of arguments that blame the victims of poverty for their own misfortune.[28]

Reflective of neoliberalism and market-driven interventions, the exergame is yet another instance where the state abdicates responsibility, placing the burden of transformation on individuals even as these same individuals lack equal access to the necessary tools.

The Biggest Loser and the countless other games ignore and erase obstacles, reducing health and nutrition to choice. The world once again matters. Likewise, all of these games are costly, further illustrating how virtual reality never exists apart from reality. So while exergames illustrate the potential to counter inequalities, to combat health disparities, and the illusive remedies for disadvantaged communities, the same injustices that perpetuate these health inequalities also impact the availability of these games. The need for large amounts of space, often unavailable for families living in small substandard homes and apartments, illustrates the ways in which the real world intersects and impacts virtual exercise. And where the homeless are concerned, even virtual exercise is not an option. The popularity and interest in reality shows and exergaming, as a nonstate capitalist intervention against a national health crisis, reflect the extent of America's collective weight issues. The numbers are telling: 34 percent of adults are obese, with another 34 percent in the overweight category. For children and adolescents, things are equally troubling, with 18 percent of kids ages twelve to nineteen and 20 percent of those ages six to eleven defined as overweight.

This issue and the lack of structural attention are particularly acute within the African American and Latino communities. Nationally, 38.2 and 35.9 percent of African American and Latino youth, respectively, ages two to nineteen, are obese and overweight, compared to 29.3 percent of whites within this same age group. In nine states, adult obesity for African Americans exceeds 40 percent, with that number between 35 and 39.99 percent for thirty-four states. "Strikingly, the link between race, poverty and obesity is most acute in the South, our nation's most impoverished region," writes Angela Glover Blackwell. "In Mississippi, which has an African American population of more than 37 percent and is the poorest state in the country, the obesity rate is the highest of any state, as is the proportion of obese children ages 10–17."[29] Worse yet, despite attention and a national discourse, not to mention the increased popularity of exergames, the obesity issue does not seem to be getting better, especially when we look within (poor) communities of color. Between 1986 and 1998, childhood obesity rates within the African American and Latino communities increased by almost 120 percent, whereas whites saw an increase of only 50 percent over this same period.

Much of the public discourse, when it acknowledges the varied rates of obesity, tends to focus on cultural differences, food choices, and varied definitions of body image with very limited discussion of the structural inequalities. The environmental barriers many youth of color face include unsafe neighborhoods, lack of open spaces, and fear of crime and violence. The psychological barriers include overcoming low social expectations for physical activity in this population that translate into low confidence and motivation to be active.[30] Exergames are incapable of addressing many of the structural and societal causes of the obesity epidemic and the ways in which inequalities manifest themselves. Yet the marketing of the games, the policy incorporation of games within public schools, and even the research itself celebrate the games as a powerful intervention against the forces that have produced high rates of obesity, diabetes, and other health-related disease. Reducing these issues to choice and individual behavior, the exergame movement treats the symptoms rather than the cause of the obesity epidemic and its particularly dangerous impact on the poor and on communities of color.

Whether talking about fear of crime, the absence of parks, or the lack of trails and walkable sidewalks, research has found a strong correlation between race, class, and opportunities of exercise. Charlotte A. Pratt emphasizes the structural and policy context for understanding the obesity epidemic and the hard-hitting impact on "low-socioeconomic and ethnic minority populations." Pushing the conversation away from individual pathology, choice, and cultural calls to "get people moving," Dr. Pratt highlights the potential and limitations of exergaming: "The Institute of Medicine (IOM) identified policy and environmental changes as seminal strategies for controlling the childhood obesity epidemic, and concluded that children and youth should be provided safe places to play, opportunities for regular physical activity, and support of their families' efforts to integrate physical activity into their daily routines. Unfortunately for children and families in disadvantaged and low-income communities, such opportunities are limited." The power of exergames stems, in part, from the limited opportunities to engage in physical activity. Yet the power of these games, as a bridge, as a mechanism to encourage physical activity or participation in physical education, after-school sports, and recreation, is simultaneously limited by these same conditions. Pratt concludes: "Taken together, these studies illustrate the powerful influences of the built environment in shaping the physical activity levels of diverse population groups, providing evidence that policy-based approaches are needed to curtail the childhood obesity epidemic, particularly in low-income populations and communities of color."[31]

While exergames provide a virtual way for gamers, irrespective of class, race, or geography, to burn similar calories, it doesn't mediate the conditions that provide health inequalities. It does not address nor it does challenge the prevalence of unhealthy foods, whether fast food restaurants or highly processed corn syrup–based "food." These games do not transport players to places where nutritious food is available or where safe parks are in great supply. Therefore, virtual efforts to encourage a healthy lifestyle, to produce new identities that will embrace healthier life choices, cannot counteract the stratification and inequality in society. The Wii can provide exercise, but in a world where we are not all the same, where we all do not have the same options and opportunities, the Wii alone is just not enough.

NOTES

1. Peter Liu, "Can Wii Games Replace Regular Exercise?," n.d., http://www.thesoko.com/thesoko/article2837.html; Julie Saccone, "Is the Wii Fit Better than Regular Exercise?," May 26, 2011, http://www.livestrong.com/article/390867-is-the-wii-fit-better-than-regular-exercise/#ixzz2JCK71Uia.

2. Saccone, "Is the Wii Fit Better than Regular Exercise?"

3. Margaret Morse, "What Do Cyborgs Eat? Oral Logic in an Information Society," *Discourse* 16, no. 3 (1994): 86; Deborah Lupton, "The Embodied Computer/User," in *Cyberspace/Cyberbodies/Cyberpunk: Cultures of Technological Embodiment*, edited by Mike Featherstone and Roger Burrows (London: Sage, 1995), 100, quoted in Micheal Sean Bolton, "Crossing Over: (Dis)Embodied Identity in Cyberspace," http://www.gradnet.de/papers/papers2004/bolton0410ng.html.

4. McKenzie Wark, *Virtual Geography: Living with Global Media Events* (Bloomington: Indiana University Press, 1994), vii.

5. Elizabeth A. Vandewater, Mi-suk Shim, and Allison G. Caplovitz, "Linking Obesity and Activity Level with Children's Television and Video Game Use," *Journal of Adolescence* 27, no. 1 (2004): 71–85.

6. "Interesting Statistics about Video Games," n.d., http://www.diyfather.com/content/Interesting_Statistics_About_Video_Games.

7. Seth Schiesel, "P.E. Classes Turn to Video Game That Works Legs," *New York Times*, April 30, 2007, http://www.nytimes.com/2007/04/30/health/30exer.html.

8. Stephen Totilo, "West Virginia Adds 'Dance Dance Revolution' to Gym Class: All of State's Schools Will Begin Using Game over the Next Two Years," January 25, 2006, http://www.mtv.com/news/articles/1521605/dance-dance-revolution-added-gym-classes.jhtml?headlines=true.

9. Quoted in Annabelle Robertson, "Can You Really Get Fit with Wii Exercise Games?," http://www.webmd.com/fitness-exercise/features/can-you-get-really-fit-with-wii-exercise-games.

10. Nanci Hellmich, "Video Games Help Schools Get Kids Moving, Exercising More," *USA Today*, October 11, 2010, http://www.usatoday.com/yourlife/fitness/2010-10-11-justdance11_CV_N.htm.

11. Donnelly quoted in Robertson, "Can You Really Get Fit with Wii Exercise Games?"; Torchia quoted in Charles Nunnmaker, "Wii Fit Becoming Out of Shape with Weight System," http://tigerweekly.com/article/03-04-2009/10520.

12. Amanda J. Daley, "Can Exergaming Contribute to Improving Physical Activity Levels and Health Outcomes in Children?," *Pediatrics* 124, no. 2 (2009), http://

pediatrics.aappublications.org/cgi /content/abstract/124/2/763.

13. Lee Graves et al., "Energy Expenditure in Adolescents Playing New Generation Computer Games," *BMJ* (December 2007): 1283.

14. Mark Anders, "As Good as the Real Thing," http://www.acefitness.org/getfit /studies/WiiStudy.pdf.

15. Lorraine Lanningham-Foster, "Activity-Promoting Video Games and Increased Energy Expenditure," *Journal of Pediatrics* 154, no. 6 (2009): 819–823; "Mayo Clinic Study Endorses Concept behind Nintendo's Wii," January 5, 2007, http://www.consumeraffairs.com/news04 /2007/01/nintendo_mayo.html#ixzz0 PrmvDSZa.

16. Bruce W. Bailey and Kyle McInnis, "Energy Cost of Exergaming: A Comparison of the Energy Cost of 6 Forms of Exergaming," *Archives of Pediatric and Adolescent Medicine* 165, no. 7 (2011): 597–602.

17. Jennifer Farish, "Can Nintendo Wii Game Consoles Improve Family Fitness?," March 7, 2011, http://www.medpagetoday. com/Pediatrics/GeneralPediatrics/25226.

18. University of Mississippi, "Wii Fit May Not Help Families Get Fit," *Science-Daily*, December 25, 2009, http://www .sciencedaily.com /releases/2009/12 /091218125110.htm.

19. Liu, "Can Wii Games Replace Regular Exercise?"

20. J. C. Nitz et al., "Is the Wii Fit a New-Generation Tool for Improving Balance, Health, and Well-Being? A Pilot Study," *Climacteric* 5 (2010): 487–491.

21. Saccone, "Is the Wii Fit Better than Regular Exercise?"

22. "*Biggest Loser* Wii Game Review," http://www.dietsinreview.com/diets /biggest-loser-wii-game/.

23. Ibid.

24. http://www.wiitalk.co.uk/forums /wii-general-discussion/14866 -demographics.html.

25. Hellmich, "Video Games Help Schools Get Kids Moving."

26. Sophie Borland, "School PE Lesson That Has Turned into a Wii Class," November 6, 2009, http://www.dailymail .co.uk/sciencetech/article-1225478 /School-PE-lesson-turned-Wii-class.html.

27. Aaron Barnhart, "Reality TV Cashing in on Obesity," March 8, 2011, http:// www.thespec.com/whatson/article /497873 – reality-tv-cashing-in-on-obesity.

28. David Grazian, "Neoliberalism and the Realities of Reality Television," *Contexts* 9, no. 2 (2010): 68–71, http://works .bepress.com/david_grazian/9.

29. Angela Glover, "People of Color at the Heart of America's Obesity Crisis," *Milwaukie Community Journal,* July 8, 2010, http://www.communityjournal.net /guest-editorial-people-of-color-at-the -heart-of-america%E2%80%99s-obesity -crisis-4/.

30. G. X. Ayala et al., "Away-from-Home Food Intake and Risk for Obesity: Examining the Influence of Context," *Obesity Research* 16, no. 5 (2008): 1002–1008.

31. Charlotte A. Pratt, "Findings from the 2007 Active Living Research Conference," *American Journal of Preventive Medicine* 34, no. 4 (2008): 366.

Contributors

ANDREW BAERG directs the Communication Program at the University of Houston–Victoria. His scholarship on digital gaming has appeared in the *Sociology of Sport Journal, Journal of Communication Studies, International Journal of Sport Communication,* and *Communication and Sport.*

MEREDITH M. BAGLEY is Assistant Professor in the Communication Studies Department at the University of Alabama, where she is also a member of the Sports Communication Program. Her work has appeared in the *Journal of Women, Politics,* and *Public Policy.*

ROBERT ALAN BROOKEY is Professor of Telecommunications at Ball State University, where he also serves as the Director of Graduate Studies for the MA Program in Digital Storytelling. His research addresses the political economy of video games. He is the author of *Hollywood Gamers: Digital Convergence in the Film and Video Game Industries* (2010). His work has also appeared in *Critical Studies in Media Communication, Games and Culture,* and *Convergence.*

MICHAEL L. BUTTERWORTH is Director of the School of Communication Studies and Associate Professor of Communication at Ohio University. He is the author of *Baseball and Rhetorics of Purity: The National Pastime and American Identity during the War on Terror* (2010) and (with Andrew Billings and Paul Turman) *Communication and*

Sport: Surveying the Field (2014). He is also founding Executive Director of the International Association for Communication and Sport.

PERRI CAMPBELL is Alfred Deakin Postdoctoral Research Fellow in the School of Education at Deakin University. Her work has appeared in the *Journal of Youth Studies* and *Critical Sociology*.

STEVEN CONWAY is Lecturer in Games and Interactivity and a member of the Faculty of Life and Social Science at the Swinburne University of Technology. His work has appeared in *Convergence, Eludamos,* and the *Journal of Gaming and Virtual Worlds*.

CORY HILLMAN is Lecturer in the Department of Communication and Dramatic Arts at Central Michigan University.

LUKE HOWIE is Senior Lecturer in the School of Social Sciences at Monash University, where he is Deputy Director of the Global Terrorism Research Center. He has authored *Terror on the Screen* (2011) and *Witnesses to Terror* (2012). Recently, he has turned his attentions to the sociology of sports media.

DAVID J. LEONARD is Associate Professor and Chair in the Department Critical Culture, Gender, and Race Studies at Washington State University. He is the author of *Screens Fade to Black: Contemporary African American Cinema* (2006) and *After Artest: The NBA and the Assault on Blackness* (2012).

MICHAEL Z. NEWMAN is Associate Professor in the Department of Journalism, Advertising, and Media Studies at the University of Wisconsin–Milwaukee. He is the author of *Indie: An American Film Culture* (2011) and *Video Revolutions: On the History of a Medium* (2014) and the co-author of *Legitimating Television: Media Convergence and Cultural Status* (2012). He is working on a cultural history of early video games.

THOMAS P. OATES is Assistant Professor of American Studies and Journalism and Mass Communication at the University of Iowa. His work on sports, media, and contemporary culture has appeared in *Communication and Critical/Cultural Studies* and the *Sociology of Sport Journal*. He is editor (with Zack Furness) of *The NFL: Critical and Cultural Perspectives*.

THOMAS G. POWER is Professor and the Chair of the Department of Human Development at Washington State University. He is the author of the book *Play and Exploration of Children and Animals* (2000), and he has published more than eighty journal articles and book chapters.

RENEE M. POWERS is a doctoral student in Communication at the University of Illinois–Chicago. Her work can be found in *Women and Language* and *Women's Studies in Communication*.

IAN SUMMERS is a doctoral student at the University of Utah.

SARAH ULLRICH-FRENCH is Associate Professor in the Department of Educational Leadership and Counseling Psychology. Her previous work has appeared in *Research Quarterly for Exercise and Sport, Psychology and Health,* and the *Journal of Sport and Exercise Psychology*.

GERALD VOORHEES is Assistant Professor of Drama and Speech Communication at the University of Waterloo. His scholarship on new media and identity has appeared in *Games and Culture* and *Game Studies: The International Journal of Computer Game Research*. He is the co-editor of Bloomsbury's Approaches to Digital Games Studies book series.

Index